Fragrant

ALSO BY MANDY AFTEL

Death of a Rolling Stone: The Brian Jones Story

When Talk Is Not Cheap: Or, How to Find the Right Therapist
When You Don't Know Where to Begin
(with Robin Tolmach Lakoff)

The Story of Your Life: Becoming the Author of Your Experience

Essence and Alchemy: A Natural History of Perfume

Aroma: The Magic of Essential Oils in Food and Fragrance
(with Daniel Patterson)

Scents and Sensibilities: Creating Solid Perfumes for Well-Being

An engraving from the 1576 English edition of The New Jewell of Health *symbolizes the art of distillation, which was used to extract essential oils from fruits, flowers, and other materials.*

Fragrant

THE SECRET LIFE OF SCENT

MANDY AFTEL

RIVERHEAD BOOKS

New York

2014

RIVERHEAD BOOKS
Published by the Penguin Group
Penguin Group (USA) LLC
375 Hudson Street
New York, New York 10014

USA · Canada · UK · Ireland · Australia
New Zealand · India · South Africa · China

penguin.com
A Penguin Random House Company

Library of Congress Cataloging-in-Publication Data

Aftel, Mandy.
Fragrant : the secret life of scent / Mandy Aftel.
p.   cm.
ISBN 978-1-59463-141-2
1. Smell.   2. Odors.   3. Cinnamon.   4. Mints (Plants).
5. Ambergris.   6. Frankincense.   7. Jasmine.   I. Title.
QP458.A34   2014   2014018554
612.8'6—dc23

Printed in the United States of America
1   3   5   7   9   10   8   6   4   2

BOOK DESIGN BY NICOLE LAROCHE

To Foster, my Philemon, from your Baucis

Aesthetics cures us of anesthesia. It awakens us.

—MICHEL SERRES, *The Five Senses*

CONTENTS

Fragrant

The "Boat of Foolish Smells," in a caricature engraving published in Josse Bade's 1502 edition of La Nef des Folles (The Ship of Fools).

CHAPTER ONE

A NEW NOSE

Odors have a power of persuasion stronger than that of words, appearances, emotions, or will. The persuasive power of an odor cannot be fended off, it enters into us like breath into our lungs, it fills us up, imbues us totally. There is no remedy for it.

—PATRICK SÜSKIND, *Perfume*

gyptians emptied corpses of their organs and filled the cavities with aromatics to prepare them for the afterlife. Romans splashed doves with rose water and set them loose in banquet halls to scent the air. Marie Antoinette employed her own perfumer, Jean-Louis Fargeon, who created bespoke perfumes to match the queen's many moods. People have feasted on aromatic materials, scented temples with them, offered them to guests. Whatever the vehicle—flowers or food, incense or perfume—people in every time and place have gone out of their way to exercise and indulge the sense of smell. Why? Because no other sense makes us feel so fully alive, so truly human, so deeply, unconsciously, and immediately connected with our memories and experiences. No other sense so *moves* us.

As an artisanal perfumer who works with extraordinary aromatic ingredients from all over the world, I venture deep into the fragrant world every day. And one of my greatest joys is bringing other people there, too, and watching as they immerse themselves in the experience. Scent is fun, sexy, visceral, transporting: it reminds us who we are and connects us to one another and to the natural world. Of all the senses, the sense of smell is the one that reaches most readily beyond us, even as it most powerfully taps the wellsprings of our inmost selves. It has an unparalleled capacity to wake us up, to make us fully human.

In *The Picture of Dorian Gray*, Oscar Wilde portrays the deep, instinctive connection between scent and our unconscious thoughts and emotions:

> And so he would now study perfumes and the secrets of their manufacture, distilling heavily scented oils and burning odorous gums from the East. He saw that there was no mood of the mind that had not its counterpart in the sensuous life, and set himself to discover their true relations, wondering what there was in frankincense that made one mystical, and in ambergris that stirred one's passions, and in violets that woke the memory of dead romances, and in musk that troubled the brain, and in champak that stained the imagination; and seeking often to elaborate a real psychology of perfumes, and to estimate the several influences of sweet-smelling roots and scented, pollen-laden flowers; of aromatic balms and of dark and fragrant woods; of spikenard, that sickens; of hovenia, that makes men mad; and of aloes, that are said to be able to expel melancholy from the soul.

Wilde knows that aromas can take us anywhere, that they are a magic carpet we can ride to hidden worlds, not only to other times and places but deep within ourselves, beneath the surface of daily life. We close our eyes—we do this instinctively before we inhale a scent, as if preparing for the internal journey—and before we even consciously recognize what we're smelling, we are carried away without our consent. Or we are stopped in our tracks, brought entirely into the present moment. (The next time you catch a whiff of skunk, try thinking about anything else at all.)

I'll never forget the first time I smelled the intense aromatic essences that are the perfumer's palette. I had signed up for a perfume class at a nearby aromatherapy studio. The teacher laid out many small bottles of naturally derived botanical essences for us to compose with: oakmoss, angelica, jasmine, frankincense, patchouli, kewda, sweet orange, lime. I leaned over to smell each one, amazed at how rich and complex and singular and stinky and alive they were, how *transporting*. As I took in the oils, in all their gorgeous diversity, it was as if a mirrored sensation were occurring inside me; I felt as if I were becoming one with the oils, as if they were entering me. I couldn't tell where I left off and they began. I couldn't and didn't want to find the words to describe them; I just felt radiant and alive—as radiant and alive as they were. I fell in love immediately.

By then I'd already lived for more than twenty years in Berkeley, California, where I had a thriving psychotherapy practice and had written a few books. The city itself had had a profound influence on me from the moment I arrived there, after an upbringing in Detroit. In Berkeley, I found the bohemian ambience I had longed for, an energy that was palpable in the streets, in the restaurants and cafés and shops. It felt as if behind every Arts and Crafts façade there were people making pottery or jewelry, writing books,

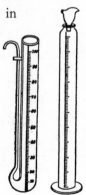

doing improv, inventing new recipes, collaborating in a kind of rampant cross-fertilization of creativity.

As it happened, I moved into an Arts and Crafts house right behind the restaurant Chez Panisse, where Alice Waters had just begun to spread the gospel of locavorism. Three houses down was the original Peet's Coffee, where Mr. Peet himself roasted the beans. My block was redolent with the smells of fresh coffee and of vegetables roasting in Chez Panisse's wood-fired oven. In Detroit, front yards had been clipped, manicured, rolled-lawn affairs, but here in Berkeley people's front yards overflowed with casual cottage gardens of fresh herbs and heritage roses, fruit trees in bloom, jasmine and wisteria climbing from basement to attic. I had never seen such a gift to the street! Despite Berkeley's reputation as the epicenter of the counterculture, the aesthetic it was steeped in was simple, almost Old World. It spoke to me, and it played a great role in shaping my own aesthetic. Working with the best ingredients, doing only what needed to be done and no more—this became my creative mantra.

It was a mantra that guided me through several career turns, fueled by the Berkeley milieu. As a weaver I worked with a range of natural materials—horsehair, goat hair, wool, silk—learning to spin them into yarn, and dyeing them with herbs and lichens I gathered. I developed an appreciation for the ways that raw materials were grown, processed, and used all over the world. Later I trained to be a therapist and focused my practice on artists and writers, to whose brilliance and creative energy I was drawn. Researching a book I was writing on Brian Jones, founding member of the Rolling Stones, I was drawn to his fascination with the costumes and music of other cultures, and to the way he embodied the conviction that anything is possible and that creativity is what life is all about.

I wrote another book, *The Story of Your Life*, marrying my fascination with plot and narrative to what I had learned about character and transformation through my years as a therapist. Then I decided to write a novel, with a perfumer as my protagonist. I knew nothing about the craft of perfumery, but its aura had allure. I signed up for that perfume class, little dreaming that in the process of researching my novel I was about to discover my true calling and become the artist I had only planned to write about.

Not only did I fall in love with essential oils, but I discovered that I had a knack for blending them. Just as sometimes you meet someone it seems you've known forever, the essences, with their distinct personalities, had a mysterious familiarity to me. I could appreciate their textures and shapes almost instinctively, like a language in which I was already fluent. As I started making perfumes, I could identify where and how I'd made mistakes, and in correcting them I learned so much about the way each essence interacted with others. I set up my own perfume business, a true cottage industry. Gradually I developed a following for my artisanal fragrances, developed from entirely natural materials that I sourced from all over the world.

At the same time, I immersed myself in the rich history of scent, acquiring more than two hundred rare and antique books on perfumery, one book leading me to the next as I fell under the spell of their charm, beauty, and eccentricity. Discovering the universe they contained was like being the first explorer in a cave that harbored the unsullied pottery and intact arrowheads of a lost civilization. In the stories of perfume, one could relive the world being discovered, retrace the footsteps of the people who came upon spices in faraway places and learned to extract the aromatic oils from exotic plants. In their intricate woodcuts and engravings, old distilling apparatuses looked like crosses between lab equipment and the tools of witchcraft. It wasn't just the history of perfume I was discovering. I had entered a heady world in which perfume commingled with medicine, science, alchemy, cooking, mysticism, cosmetics, and craft. There lay the richer, more synesthetic sensibility of a bygone time, a beautiful and mysterious universe of magical things jumbled together.

Eventually I did write another book of my own, not a novel but *Essence and Alchemy: A Natural History of Perfume*, which introduced perfumers and "perfumistas" to the sensual history of natural aromatics and the building blocks of creating perfume with them. The nascent natural/artisanal perfumery movement embraced the book, and I began to teach the art of perfumery as well as to practice it. I also began to confer and collaborate with chefs and mixologists who were coming at natural essences via food and drink. I found myself in the vanguard of a surge of interest in scent as a key component of flavor, an exciting new arena I explored in a cookbook I coauthored with Michelin two-star chef Daniel Patterson.

In my nearly two decades of peregrinations through the world of scent, the wonder of encountering amazing new fragrances has never left me. And everywhere I've gone, I've had the joy of giving other people that experience. Sometimes they try to head me off, claiming they are "not into perfume." Instinctively repelled by their exposure to an olfactory diet that's the equivalent of fast food—the assaultive, artificial scents that saturate what we eat, our cleaning products, mass-produced perfume, the very environment—they've come to believe they have no appetite for scent itself. Watching them discover authentic aromas and their sensual pleasures is profoundly thrilling, like watching a starving person feast on a delicious meal. It's these experiences of reawakening people to scent that led to this book. I wanted to write next for everyone, not just for the perfumers and perfumistas—though I think they too will find new information and inspiration here, as their passionate interest in high-quality perfumes has fueled the trend toward niche brands and small-batch perfumery.

As Michael Taylor observes in *Rembrandt's Nose*, that great painter intuitively understood that the nose was the key to understanding a person's face. He painted noses that "possess a will of their own."

They have their own inclinations and seem to obey their own promptings rather than the laws of objective resemblance. They are long and slender, flat and squat, smooth or wrinkled, bony or fleshy, dainty or gross, pitted, scarred, inflamed, unblemished—less, one feels, for reasons of fidelity to the sitter than for reasons dictated by the artist. . . . He rendered the complexion of a nose with the same fastidiousness that he brought to paraphrasing the sheen of velvet or fur. In his portraits and self-portraits, he angles the sitter's face in such a way that the ridge of the nose nearly always forms the line of demarcation between brightly illuminated and shadowy areas. A Rembrandt face is a face partially eclipsed; and the nose, bright and obvious, thrusting into the middle of halftones, serves to focus the viewer's attention upon, and to dramatize, the division between a flood of light—an overwhelming clarity—and a broody duskiness.

I feel a bit about the nose as Rembrandt evidently did. The nose is idiosyncratically central not only to our sense of smell but to our sense of *who we are*, in our most primal appetites. For the idea of appetite pertains to food as well as to all the sensual and spiritual experiences that drive us, give us pleasure, make us feel more alive in the moment. Scent is a portal to these basic human appetites—for the far-off, the familiar, the transcendent, the strange, and the beautiful—that have motivated us since the origins of our species.

As I researched and thought about the deep ways that perfume touches our most primal selves and the collective self of our species, I realized that I had the makings of an adventure story of sorts, an entrée to writing about scent as a series of excursions into the fragrant world that I think will return you more awake and alive, more profoundly able to "smell the roses."

To narrate it I chose five landmark scents—think of them as five rock stars of the aromatic world. Each represents a key story line in the history of scent, intricately bound up in its adventures and intrigues and moments of discovery. Each also represents a class of material from which are derived the ingredients essential to the art of perfumery (and also to the art of cooking and flavoring). And each touches on—and stirs—one of the basic appetites that define us.

- CINNAMON, among the spices, attests to our appetite for adventure—the pursuit of the exotic and the luxurious. It's a link to an era of risk and discovery, a time when not all was mapped and known.
- MINT, among the herbs, speaks to the lure of home—our hunger for the familiar, the authentic, the native. Mint is the quintessential American plant and a link to the "weird old America" of a bygone era that lives on in the national character, beneath the topsoil of modernity. But it is also indigenous to almost everywhere, a universal symbol of welcome.
- FRANKINCENSE, one of the deeply aromatic resins traditionally used in incense, represents our inclination to the spiritual. The trees these resins come from are like a mani-

festation of the urge they embody—to reach beyond, to transcend our temporal being by means of ritual and out-of-body experience.

- AMBERGRIS, an animal essence, is testament to our unquenchable curiosity—our fascination with the unusual, the strange, the wondrous, the "other."

- JASMINE, among the floral fragrances, stands for our yearning for beauty, for an aesthetic that embraces the evanescence of our existence as well as what endures—an aesthetic embodied in the Japanese design philosophy of wabi-sabi.

In *A Natural History of the Senses*, Diane Ackerman reminds us that our prehuman selves started in the ocean, relying on the nose to seek out food and identify enemies. "In our early, fishier version of humankind . . . smell was the first of our senses," she writes, and cognition evolved from it: "Our cerebral hemispheres were originally buds from the olfactory stalks. We *think* because we *smelled*." Scent has helped each of us, since we were babies, to recognize what is of us (familiar) and not of us (other). Babies learn to recognize their mothers by smell—so important when you spend a lot of time with your eyes closed!

On a molecular level, scent actually touches your nerves. In order for you to perceive a distinct smell, some volatile constituent of the odoriferous substance must vaporize, waft deep into your nose, and settle on your olfactory receptors. Vision and hearing, by contrast, depend only on the perception of light waves or sound waves that emanate from or are reflected by the object in question; in a sense they only describe the object in question, while smell involves physical contact with its emanation. If someone places a batch of just-baked chocolate-chip cookies on the counter, their tantalizing aroma enters your nose and triggers a sensory response, touching you even before you can reach out your hand to grab one, even if you leave without taking a bite.

A single smell instantly brings forth a cascade of memories. Experiences and emotions get primally attached to smells, and those associations of memory and feeling are immediately triggered when the smell is next encountered—even if "next" is twenty years later. We eventually put words to these associations, to make them into stories—you associate the smell of pipe tobacco with being in your grandfather's study—but they start as a visceral connection, fundamental to our human DNA. These primal associations make scent incredibly personal and specific, its associative fingerprint unique to each individual. As the philosopher Michel Serres writes in *The Five Senses*:

> Smell seems to be the sense of singularity. Forms reappear, invariant or recurrent, harmonies are transformed, stable across variations, specificity is countersigned by aroma. With our eyes closed, our ears stopped, feet and hands bound, lips sealed, we can still identify, years later and from a thousand other smells, the undergrowth of such and such a place in a particular season at sunset, just before a rain storm, or the room where feed corn was kept, or cooked prunes in September, or a woman.

And yet the power of scent also derives from its ubiquity. At a given moment, it seems to be pervasive, everywhere at once. Smells emanate from unseen sources, often from many sources at once. They have no GPS tracking built in to tell us where they come from. They are like the ultimate superhero— invisible, untraceable, and sometimes overwhelming. Noxious smells seem like the very vehicles of contagion. Before the discovery of germs, writes William Miller in *The Anatomy of Disgust*, people thought that unpleasant odors spread disease while pleasant smells cured it. The associations  persisted beyond the advent of germ theory, in fact—which is why cleaning products "must have a smell that accords with our beliefs about the smell of asepsis," Miller observes, even though the smell has nothing to do with their cleansing properties.

> *Perfume would seem to be one of the elements, one of the original secrets of the universe. How it gets into flowers, and certain uncouth creatures, as for instance, ambergris in the whale, or civet in the civet cat of Abyssinia, or musk once more in the Florida alligator, is a hidden process of the divine chemistry, and why it affects us as it does no philosopher has yet explained. Literally, it belongs to those invisible powers whose influence is incalculable, and as yet unknowable.*
>
> —RICHARD LE GALLIENNE, *The Romance of Perfume*

Notwithstanding the inroads that scientific knowledge has made, scent remains one of the most accessible yet irreducible experiences of magic that we have. Indeed, entering the world of fragrance is like falling through the looking glass and finding on the other side an everyday miracle, a mystery, a source of wonder. It is a truly transformative experience, and one that I am passionate to share.

So smell has a potent pull on human beings: it has a prehistory with our species. But the unique power of scent has also given it a rich *history* with humanity. Lands have been discovered and conquered for the sake of perfumed materials; with them the most elusive lovers have been seduced, the most implacable gods worshipped. They fuel the pursuit of the extraordinary and return us to the comfort of home. Aromas come freighted with the stories of where they've been and what's been done and dared in their name, as surely as do legends from far away and long ago. Through them we feel our place in the long narrative that is history: they bear the imprint of our cultural DNA.

Evidence of the vibrant aromatic life of bygone times is as rich and compelling as any archaeological find. The beliefs of the ancient Romans emanated from a scent-steeped existence: "Spiced wine mingled with frankincense could drive elephants mad, even if militarily trained; wild beasts could be lulled to sleep with flower petals and aromatics," writes Susan Harvey in *Scenting Salvation*. "Perfume was lethal to vultures, a bird that lived on putrid carrion, but could tame an entire flock of doves. Panthers could emit a sweet scent by which to beguile their prey. The ancient industry of aromatics was less an attempt to harness the powers of smells than to participate in their abundance."

Life in earlier epochs was imbued with a full panoply of aromas, including the impolite smells of humans and animals: urine, feces, rotting food, smoke, sweat, illness, and death. Without modern means of hygiene and sanitation, odors—both good and bad—were irrepressible, pulsating with life, and intensity was the norm rather than an occasion for disgust. As Paul Freedman observes in *Out of the East*, "It is precisely because of this inevitable familiarity with awful odors that people in premodern societies were entranced with beautiful smells. They experienced a wider spectrum of olfactory sensations than we are familiar with, both good and bad. What tended to be missing was the neutral non-smell of modernity."

There are signs that we are all tiring of this neutral non-smell of modernity. Since I published *Essence and Alchemy*, a new world of appreciation has sprung up around smell, much the way the world began to wake up to food a decade or two ago. People inspired by Alice Waters, Michael Pollan, and other apostles of the Slow Food/locavore movement have taken an interest in the provenance, quality, and variety of what they consume, not only for health reasons but to have a pleasurable, authentic connection to a universal experience. Perfume, which used to be covered by the press only twice a year and solely in women's fashion magazines, is now covered by the minute, by literally hundreds of perfume, fashion, and makeup blogs that have mushroomed on the Internet. The perfumistas who follow these blogs do not blindly buy the most heavily advertised perfumes but read, talk, meet, and breathe scent, developing and voicing their own sophisticated opinions. The ranks of those interested in wearing, using, and knowing about scent are growing exponentially. As I meet and talk and tweet and Facebook with them, I have been privileged to witness, again and again, the profound power of the sense of smell, the longings and impulses it stirs, the desire and appreciation for a richer connection with life.

"I remember with gratitude the moment when a great wine gave me a new mouth," writes Michel Serres on savoring a bottle of 1947 Château d'Yquem Sauternes. I want to give people a new nose, to introduce them to the heady sensuality of a fully engaged sense of smell. I want them to reawaken to smell, the way they have begun to reawaken to taste after a decades-long slumber. Often the path to awakening scent comes through taste, its closest sibling among the senses. In fact, "taste" is really a product of the two senses—the sensation of taste on the tongue plus smell in the nose—so the reawakening to food has really been in part a reawakening to smell. By focusing on scent, I want to make us aware of its primacy and power, but the two remain entwined, to the enrichment of both. As with taste, the pleasures of fragrance are transitory and evanescent, but they offer an incomparable, unadulterated experience of joy—they speak to first principles of pleasure, and for the moment that we experience them, our appetites are stronger than the fragility of life.

In *Worlds of Sense*, Constance Classen writes about the Ongee, an isolated tribe of hunter-gatherers living on a remote island off the coast of India, who are the most aromacentric community I have ever heard of. For the Ongee, smell is a source of personal identity, a system of medicine and communication:

When an Ongee wishes to refer to "me," he or she puts a finger to the tip of his or her nose. . . . The most concentrated form of odour according to the Ongee are bones, believed to be solid smell. The Ongee thus say "smells are contained in everybody like tubers are contained in the ground." An inner spirit is said to reside within the bones of living beings. While one is sleeping, this internal spirit gathers all the odours one has scattered during the day and returns them to the body, making continued life possible.

The Ongee hold illness to result from either an excess or a loss of odour. . . . The basic treatment for an excess of odour consists of warming up the patient in order to "melt" the solidified smell. A loss of odour, in turn, is treated by painting the patient with white clay to induce the sensation of coolness and restrict the flow of odour from the body. . . . The concern of the Ongees to maintain a healthy state of olfactory equilibrium is expressed in their forms of greeting. The Ongee equivalent of "how are you?" is . . . "how is your nose?" Or literally, "when/why/where is the nose to be?" . . . Death is explained by the Ongee as the loss of one's personal odour. They believe that they kill animals they hunt by letting out all of their smell and that they themselves are hunted by spirits, called *tomya*, who kill them by absorbing *their* odours. Birth, in turn, is caused by a woman's consuming food in which a hungry spirit is feeding. . . .

A newborn has soft bones and no teeth, hence possesses little odour. On growing up, a child develops the condensed odour contained in hard bones and teeth. In old age a person loses odour through illness and the loss of teeth, until death reduces the person to a boneless, odourless spirit—which will eventually be born again as a human. . . .

Living in a community is believed to unite the odour of individuals and lessen their chances of being smelled out by hungry spirits. When moving as a group from place to place, for example, the Ongee are careful to step in the tracks of the person in front, as this is thought to confuse personal odours and make it difficult for a spirit to track down an individual.

We don't analyze scent; it unfolds in us. Aspects of an aroma strike us, then fade away as other notes emerge and gradually disappear. Like music, smell is an evolving experience, always in motion; the nuances of aromatic notes are experienced in transition from one to another. And as with music, the intangibility of scent allows us to experience it in a state of dreaming imagination. Smell is an invitation to a journey: it allows us to leave the ordinary course of things and go on a trip, to absent ourselves.

Not everyone has the desire or the ability to become a perfumer, any more than everyone has the desire or ability to become a chef. But I fervently believe that everyone can regain an authentic connection to the sense of smell. Ultimately my aim is not to dissect its power—not to pluck out the heart of its mystery—but rather to allow anyone with curiosity and a yearning to inhabit life fully to enter deeply into it.

With the aromas of cinnamon, mint, frankincense, ambergris, and jasmine as our spirit guides, we will explore the secret life of scent. Along the way we'll stop in at my workshop—less to make a perfumer out of you (unless you're so inspired!) than to introduce the essentials of scent literacy. In the same way that knowing a little about how to play an instrument enhances your enjoyment of music or taking an art history class makes you a more sophisticated museumgoer, understanding the basics of how scent is composed deepens your appreciation of the fragrant universe. And I'll share recipes for some easy-to-make perfumes and edible, drinkable, and useful concoctions that will reveal the imaginative possibilities of creating with—and reveling in—aroma.

HOW TO SMELL

Imagine taking the first sip of a fine wine or listening to beautiful music: most of us begin almost instinctively by closing our eyes to shut out visual distractions. Approach smell the same way. Close your eyes, and eliminate or minimize other stimulation, too—give yourself a quiet, spacious, uncluttered place to smell, free of competing odors. Bring a concentrated focus to bear on the aromatic sensation that you are receiving. This should not be done absentmindedly or passively. Push aside simple dichotomous reactions—*Yes, I like this* or *No, I don't*—and try to appreciate that a scent is something alive, vibrant and nuanced and unique. Allow your instinctive affinities to surface. We have an innate feeling for natural aromas; they take us back, unbidden, to a primordial state in which we recognize the shape, texture, and evolution of a scent. Even if we haven't encountered vetiver before, for example, we tend to "recognize" it as exotic, faraway, grassy. Let your sense of smell wander. Notice:

- the layers of smell (one-dimensional or complex and layered)
- the shape of the smell (pointed, sharp, rounded, dull)
- the memories it conjures
- the feelings it arouses

Ask yourself: If this fragrance were a color, what would it be? Allow the smell to open itself to you, and discover whatever about it is most beautiful, most remarkable, to you.

OLFACTORY FATIGUE

After I have been working with essences for a while, I inevitably get to a point of *olfactory fatigue*, when my ability to perceive aromas is saturated and therefore blunted. I'm like a cup that's full and can hold no more.

As with all our senses, overstimulation temporarily dulls our sensitivity. Scientists theorize that as our sensory systems are evolutionarily calibrated to be on the lookout for change in our environment, constant bombardment eventually stops registering as change and instead sets a new baseline. Food science writer Harold McGee explains, "Even though you're smelling a series of different things, they do have chemicals in common, and the repeated exposure to them causes our brains to tune them out. So our perception of scents becomes unbalanced and less true to each scent in its totality."

When I teach, I notice that the onset of olfactory fatigue is different for different students, even when they've been smelling the same materials. Some people seem to have a "tougher" nose than others. And it's possible to build up stamina as you work with scented materials. But sooner or later olfactory fatigue will set in with even the hardiest of us.

Coffee beans as a refresher for a tired olfactory palate are a fixture of retail perfume counters, but according to psychologist Avery Gilbert, this trick doesn't actually work. "It's all good fun and marketing, but there is not a jot of science behind it," he observes in *What the Nose Knows*. The coffee aroma provides a diversion, perhaps, but not the respite that your nose truly needs.

Here's a more effective cure for olfactory fatigue: When I notice that the essences I'm working with have begun to smell weak, I inhale three times deeply though a piece of wool. A sweater or scarf works fine;

using wool seems to be the key. As McGee theorizes, perhaps because the lanolin and other waxy components of the wool absorb and neutralize aromas, it provides the odor-free sniffing that your brain needs. Breathing through wool, he suggests, "is a more convenient version of sticking your head out the window for a minute. In each case you avoid breathing the residues of the scents you've been smelling."

WHAT ARE NATURAL ESSENCES?

You are already, and often, in the presence of natural essences—they are what make the heavenly aromas in fresh herbs and spices, citrus peel, vanilla bean. If you drag your fingernail along the skin of a lemon, the essential oil in the rind will transfer to your finger. Essential oils are where the flavors live in plants, and can be extracted from them by various means. In fact, good-quality essential oils have an aroma and flavor whose intensity exceeds that available in the plant material itself. Essential oils are not the same as vegetable oils pressed from seeds, nuts, olive flesh, or the bran of grain, all of which yield a greasy oil. "Essential oils," in fact, is a bit of a misnomer, for they contain no fats; left exposed to the air, they will not linger but will evaporate completely. In other words, they are volatile aromatic compounds whose molecules will quickly become airborne vapors that easily reach our noses. (Technically these materials usually have molecular weights of less than 300.) They are extracted and condensed into a concentrated liquid form, eminently suitable for creating fragrances and flavors, but we should bear in mind that when using these extracts, we are basically working with fumes.

The odors of plants reside in different parts of them: in the bark (cinnamon), roots (ginger), blossoms (rose, ylang-ylang), leaves (basil, tarragon, bay, marjoram, rosemary, mint, sage, thyme), seeds (cumin, coriander, nutmeg, allspice, anise, juniper, cardamom), and rind (lemon, lime, orange, grapefruit). Sometimes different parts of the same plant produce distinctly different essences. For example, the bitter orange tree yields petitgrain from its leaves and twigs; neroli and orange flower absolute from its blossoms; and bitter orange essential oil from the rind of its fruit.

A note on safety: Some natural essences have been known to cause allergic reactions; you should avoid natural essences during pregnancy. For safety guidelines with essential oils, consult Robert Tisserand's *Essential Oil Safety: A Guide for Health Care Professionals.* The International Fragrance Association (IFRA) has compiled a list of recommended guidelines for commercial perfumers, which is updated periodically. You can find it at www.ifraorg.org.

ESSENTIAL OILS Essential oils are the largest category of fragrance materials and the most widely available, thanks to the tremendous popularity of aromatherapy.

The essential oils in the rinds of citrus fruits are extracted by simple pressing. Most other plant materials are put through a process of distillation with water or steam. The method depends on the fact that many substances whose boiling points are far higher than that of water are volatilized when they are mixed with steam. The volatile substance must not be soluble in water, so that, on cooling, it separates from the watery distillate and can be preserved in a relatively pure condition. In direct distillation the plant material is in contact with the boiling water. Steam distillation is a gentler method of extracting essential oils, and more widely used.

CONCRETES AND ABSOLUTES Even steam distillation is too harsh a process for some materials—namely, flowers, whose natural perfume molecules decompose under the heat of distillation. Natural flower oils are therefore separated from fresh flowers by solvent extraction. In solvent extraction, flowers are placed on racks in a hermetically sealed container. A liquid solvent, usually hexane, is circulated over the flowers to dissolve the essential oils. Because the flowers give off a great deal of waxy material, the process yields a so-called concrete, which is semisolid. Concretes have a softness to their aroma and at the same time great staying power. Although they are not completely soluble in alcohol, they are perfect for making solid perfume. If they are infused into a liquid perfume, the insoluble dregs need to be strained after the aging process.

By dissolving the waxes with pure alcohol (ethanol) and removing other solids, a concrete can be rendered into an absolute, a highly concentrated liquid essence that is entirely alcohol-soluble. Absolutes are floral essences at their truest and most concentrated. They are much longer-lasting than essential oils and have an intensity and fineness to their aroma that are unequaled. They are the most expensive perfumery ingredients.

CO₂ EXTRACTS Carbon dioxide extraction is a relatively new method, similar to distillation. At normal temperatures carbon dioxide is usually a gas. But with enough pressure it can be turned into a liquid that acts as other liquid solvents do. Instead of water or steam, pressurized carbon dioxide is pumped into a chamber filled with plant matter, to extract the essential oil along with other substances, such as pigment and resin. This can be done at lower temperatures than with water-based distillations, around 100 degrees Fahrenheit. And in this process, unlike with liquid solvents, to separate the aromatics from the CO_2, all you have to do is bring the temperature back down to normal. Thanks to the gentle changes in heat, CO_2-extracted oils are closest in constitution to the oils as they reside in the plant itself.

NATURAL ISOLATES An essential oil is a cocktail of different kinds of aroma molecules: major constituents, minor constituents, and trace constituents. A natural isolate is one of those aromatic components separated from the others, much as an egg yolk can be separated from the whole egg to be used in a recipe. Natural isolates have allowed natural perfumers to expand their aromatic palette, especially with respect to florals, giving their fragrances a sheerer, lighter texture.

Industrial-fragrance companies frequently use single aroma constituents that they manufacture from various base-chemical materials. These are sometimes called "nature identical," but they haven't been isolated from natural materials.

ESSENTIAL OILS FOR FLAVOR

Cooks have long employed essential oils for flavor, by pounding spices with a mortar and pestle or tearing up the leaves of herbs to release their fragrance. British cookery icon Patience Gray highlights the primacy of aroma in flavor: "The secret of cooking is the release of fragrance and the art of imparting it."

Cooling Cups and Dainty Drinks, a recipe book from 1869, gives a couple of techniques for maximizing the release of essential oils from citrus peel. It can be sliced as thinly as possible ("by reason that the flavour and scent, which constitute its most valuable properties, reside in minute cells, close to the surface of the fruit, so, by slicing it very thin, the whole of the minute receptacles are cut through, and double the quantity of the oil is obtained") and mashed with sugar to a stiff paste with a mortar and pestle. Or a piece of sugar can be scraped over the outer rind of the fruit to become saturated with the essential oil. The sugared essence that results is referred to as *oleo-saccharum*.

You can easily imagine the use of such flavored sugars in custard, cookies, and cakes. In fact, there are lots of ways to cook with essential oils, adding one or more at a time to recipes. Edible essences are available (see the Sources section at the back of the book), and throughout this book you will find exercises demonstrating how to use them for flavor as well as fragrance. Making flavors with essential oils is creative, easy, and fun, and the results are delicious. It is as simple as adding a drop of ginger oil to a pot of tea, a drop of basil oil to the dressing for a tomato salad, or a drop of peppermint oil to chocolate frosting. Using a drop of essential oil can be very convenient when you just need a pinch of, say, cilantro or basil and don't have a fresh bunch on hand. Even more important, essential oils lend an intensity that can be difficult to achieve with raw ingredients and impossible with dried. However, it's important to be aware that not all essential oils are safe to ingest. You can find a list of those that the FDA considers GRAS (Generally Recognized As Safe) here: www.fda.gov/food/ingredientspackaging labeling/foodadditivesingredients/ucm091048.htm.

SHOPPING WITH YOUR NOSE

Shopping for beautiful ingredients at a farmers' market or at any quality purveyor is a wonderful opportunity to engage your sense of smell. Rip up a couple leaves of rosemary, thyme, or mint or the petals of a fragrant rose and smell your fingers. Use your olfactory faculties to find the best examples of the foods you love. I am a big tea drinker, and I take great pleasure in smelling the different varieties before I settle on one.

If I am ordering tea online, I order small samples first. When I smell them, I close my eyes and let the aromas wash over me without any analysis but with close attention to my reactions: I wait for the one that elicits a pure and enthusiastic *Yes!* Do the same when you shop—not just for tea but for melons, strawberries, apples, peaches, cheeses, coffee, spices, and more.

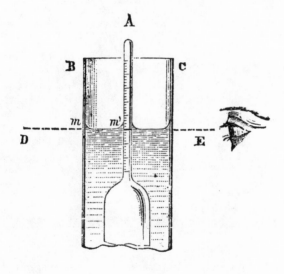

A miniature of a man wielding a stick against a cinnamologus, *a mythical bird described by Herodotus and in various bestiaries as building its nest from cinnamon, appears in a volume of writings by Peraldus (William Perault, circa 1190–1271).*

CHAPTER TWO

A TASTE FOR ADVENTURE

Cinnamon

I am the cinnamon
peeler's wife. Smell me.

—MICHAEL ONDAATJE,
"The Cinnamon Peeler"

*A*s I immersed myself in learning about the incredibly rich history of perfume, I often found myself retracing the paths of those who had gone before me, seeking out unknown, unsmelled, and untasted ingredients all over the world. Every time I came across a passage in one of my antique perfume books that introduced an unfamiliar and irresistible-sounding essence, I had to track it down so that I could start working with it; I had to get my hands on it.

It's still one of my favorite aspects of being a perfumer, searching in all corners of the globe for the most intense and distinctive scent materials, in their purest forms. I feel like an adventurous queen on the modern-day Spice Route—to me belongs the hunt. The sweet, the foul, the spicy, and the putrid—I find them all alluring. I love the complicated histories of the materials and the complex characters that make the natural perfumer's palette so vibrant. They are at once delicate and pungent, fresh and decadent, floral and fecal (yes, "fecal-floral" is actually a term of the perfumer's art—but more on that later). Even the names seduce: costus, ylang-ylang, choya loban, boronia, civet, tonka bean, champaca. Some names— ambergris, jasmine, sandalwood, frankincense, myrrh—sweep along with them associations to myth and legend, ancient history and exotic customs; they conjure up images of long desert journeys and scenes of sensual torpor. And the aromas themselves are transporting: they lift you out of the day-to-day and take you far away, both feeding and stoking the appetite for the unfamiliar, the unattainable, the indefinable.

The spices were unknown in the West until they arrived there from the most far-flung places, and then became a major driver in international trade and conquest. So as we trace the connection between fragrance and the appetite for adventure, it's the rich and varied story of spices we'll be following. And sometimes we'll let cinnamon—the first of our five rock-star fragrances—lead the way.

IN PERFUME'S WAKE

As the saying goes, the world was discovered in perfume's wake. Civilizations were built, and fortunes made and lost, in pursuit of precious aromatics, prized equally as spices and as scent ingredients. It's no mystery to me why procuring them was worth the hardships of camel caravan and cargo ship, the treacherous passage across dozens of borders and roiling seas and thousands of arduous miles, for century upon century. Or why the expansion of the spice trade across the Mediterranean was motivation enough for Alexander to found Egypt's great port city and name it for himself. It's what drove Columbus to sail the wrong way around the world, seeking a more direct route to the spices that he was certain the "Indians" he came upon possessed. It had the Dutch and the Portuguese waging war over spices for sixty years, across the Americas, Africa, India, and the Far East. It's the same reason that I myself am so turned on by them: Spices smell good! As Paul Freedman puts it in *Out of the East*, "It is impossible to overstate the importance of fragrance itself, of intoxicating and wonderful scents in a world of unpleasant odors, decay, and infirmity."

Spices and perfumes are mentioned in the records of ancient Sumer, which developed in the region of Mesopotamia around 3000 B.C. The Sumerian word for perfume is made up of the cuneiform signs representing "oil" and "sweet." In that early period, as for millennia afterward, oils were infused with spices to create perfumes.

In the third century B.C., Alexander the Great expanded the Greeks' reach from the Adriatic Sea to the Indus River and founded the port of Alexandria in order to extend the lucrative spice trade into the Mediterranean. After conquering Gaza he brought back frankincense resin with which to perfume Greek temples and, from Persia, saffron for use in infusions, rice, and even bathwater (as a curative for battle wounds). The ancient Greeks also had a predilection for complicated spice blends with long staying power that they sprinkled on bedding, clothing, and the body. "Indeed, simple, unmixed scents appear to be a comparatively modern preference," observes Edward H. Schafer in *The Golden Peaches of Samarkand*.

Along with saffron, black pepper and ginger were key players in the spice trade. From early Rome through medieval times and on through the Renaissance, strong spices were all the rage for cooking a wide variety of dishes—up until the understated richness of classical French cooking set the standard for much of the Western world in the eighteenth century, relegating such spirited seasoning to restrained doses in puddings and desserts. It used to be thought that people of earlier times used spices to retard the spoilage of meats or to mask the taste and odor of decay when it set in. But spices don't preserve meat very well, compared with salting, smoking, or pickling, and the bad taste of rotten meat cannot be masked or improved much by spices. Moreover, meat was relatively cheap, whereas spices were very expensive. The fact seems to be that people ate spicy foods because they liked the taste of spices.

Still, the idea that these scarce and precious substances came from far away gave them additional savor—and increased their value. One of the spices the Greeks prized in their blends was silphium. Around 630 B.C. they founded the colony of Cyrene, in what is now Libya, purely to exploit this particular spice, which was indigenous; it generated seven centuries of bustling trade before it became extinct as a result of overconsumption.

Because spices grew in very different climates and took a year or more to reach Western markets, they arrived in a dried form that gave little clue as to their original state. Very few Europeans had seen ginger, pepper, nutmeg, or cloves freshly harvested, let alone growing as part of a plant—another source of mystery. Herbs were a different matter. Many were native to Europe, and the old herbal guides contained accurate drawings of them. Even the leaves of plants that weren't native were not difficult to imagine in their original context. But the drawings of tropical spices from the same period are completely fanciful: they were strange, unseen. It wasn't until Marco Polo visited India and southern Asia in the thirteenth century that Europeans began to develop an idea of how tropical spices actually grew.

Moreover, spices were not only exotic-looking; they were exotic-*tasting*—pungently aromatic, even more so in their dried form than fresh. It is worth pausing for a moment to ask why this is so, from an evolutionary perspective. In contrast to the attractive aromas of flowers, for example, spice aromas generally evolved as a deterrent to predators and indeed are sometimes toxic to them. Yet the same smells that repel pests also result in the culinary, medicinal, and aesthetic uses to which humans have put spices for centuries. The key, as Harold McGee expresses brilliantly in his seminal book, *On Food and Cooking*, is intensity: "And yet humans have come to prize these weapons that are meant to repel us. What makes herbs and spices not only nontoxic and edible but delicious is a simple principle of cooking: dilution. If we bite into an intact leaf of oregano or a peppercorn, the concentrated dose of defensive chemicals overwhelms and irritates our senses; but those same chemicals diffused throughout a dish of other foods—a few milligrams in a pound or two—stimulate without overwhelming. They add flavors that our grains and meats don't have, and make those foods more complex and appealing."

THE TITAN OF THE SPICE TRADE

It was cinnamon, today a commonplace resident of practically every kitchen cabinet, that was far and away the most esteemed of the Eastern spices. In the fifth century B.C., Herodotus gave this rather fanciful account of collecting cinnamon in the wild:

The Arabians say that the dry sticks, which we call *kina-momon*, are brought to Arabia by large birds, which carry them to their nests, made of mud, on mountain precipices which no man can climb. The method invented to get the cinnamon sticks is this. People cut up the bodies of dead oxen into very large joints, and leave them on the ground near their nests. They then scatter, and the birds fly down and carry off the meat to their nests, which are too weak to bear the weight and fall to the ground. The men come and pick up the cinnamon. Acquired in this way, it is exported to other countries.

The Egyptians used cinnamon in mummification, and the Greeks and Romans used it in religious rites, adding cinnamon sticks to ritual fires to create aromatic smoke or burning it like incense. Philostratus's account from the second century B.C. describes a brew that included cinnamon, gems, and chopped snake and was supposed to impart the power to converse with animals. A Roman emperor had crowns made from cinnamon and covered with beaten gold. Pliny the Elder noted 350 grams of cinnamon as being equal in value to over five kilograms of silver. It became the titan of the spice trade, sought after not just for its culinary uses but for its ability to punctuate the shape of a perfume, especially in combination with floral ingredients.

There was a constant push along the Spice Route to discover the sources, control the trade, and even take over the production of spices, if not the lands themselves. Almost everyone was interested in a piece of the cinnamon action, and control over its supply passed through many hands over time. The Romans were initially supplied by Arab dealers, who brought the spice by overland trade routes to Alexandria. Later, Venetian traders developed a monopoly on the spice trade in Europe, but when the Mamluk sultans and the Ottoman Empire disrupted it, European traders searched for other routes of supply. Portuguese merchants finally reached the ur-source of cinnamon, Ceylon (present-day Sri Lanka), at the beginning of the sixteenth century, establishing a fort there to protect their new monopoly for the next hundred years. Then Dutch traders arrived and made a treaty with the inland Kingdom of Kandy, allowing them to take over control, and within twenty years they had expelled the Portuguese.

You can read the intense value of the spice in the intense regulation and control it generated. Once the Dutch started setting up cinnamon plantations in Ceylon, they made every attempt to restrict its cultivation to the island. Native bark collectors, understanding that their livelihood was at stake, tried to resist, but the Dutch decreed that any worker caught in what they defined as an act of rebellion—such as cutting a cinnamon shrub—would have his right hand chopped off. Furthermore, anyone who removed a cinnamon seed from the island or bought cinnamon on the black market was guilty of a crime punishable by death. This reign of terror was so successful that it resulted in an abundance of Ceylonese cinnamon—which of course lowered the spice's profitability. The Dutch responded by burning the surplus supply not only in Ceylon but even at the end of the line in Amsterdam, where the aroma of roasted cinnamon covered the city for days. The birds, however, were beyond Dutch control; they carried the now abundant cinnamon seeds and planted them in other countries, according to the legend reported by W. M. Gibbs in his landmark 1909 book *Spices and How to Know Them*. (At least that is the legend; the likelihood is that the same species was native to mainland India and Burma.)

Birds notwithstanding, cinnamon remained an exceedingly lucrative product. Gibbs reported that the owners of Ceylon's cinnamon plantations, or "gardens," lived like royalty, their homes full of carved wood that was truly worth its weight in gold. At that time true cinnamon sold for as much as a hundred dollars a pound, "while the ordinary China cassia handled by our grocers sells at wholesale at six or seven cents a pound."

The best-quality cinnamon oil opens with a candylike freshness that evolves into a tenacious aroma that is equally sweet and spicy, finishing off with a dry sweetness. The spicy heat builds gradually, with nuances of floral and clove notes. Cassia (and inferior cinnamon) hits you hard, with the aroma and taste of the little red candies known as Red Hots. Not only is the quality of cinnamon infinitely superior to that of cassia, but there is enormous variation in the quality of cinnamon itself, even within that derived from various parts of the same tree. The aromatic essence is distributed, with varying intensity, throughout the cinnamon tree, every part having a distinct flavor that can be used somehow. Any given shoot from a particular tree may yield an oil quite different in quality from that of its neighbors. Gibbs explains the ranking: "The quality of the bark depends upon its position on the branch; that from the middle is the best, that from the top second, and that from the base, which is the thicker part of the branch, the third grade."

Moreover, branches exposed to direct sunlight as they grow "have their bark more acrid and spicy than the bark of those which grow in the shade. . . . The quality is determined by the thinness of the bark—the thinner and more pliable the finer. The finest quality is smooth and somewhat shiny and of a light yellow color. The shoot bends before it breaks, and when the fracture occurs it is generally in the form of the splinter which has an agreeable, warm, aromatic taste with a slight degree of sweetness." The quality of the cinnamon also depends heavily on the growing conditions; a marshy soil produces a spongy bark so bereft of aroma that it may be useful only for adulterating more valuable grades of cinnamon.

The process of harvesting cinnamon is not much changed from that described by Gibbs a century ago. In the first stage of the harvest, the "flush" of tender shoots is cut down and left to ferment lightly indoors. The leaves are removed, and the outer bark is scraped off the twigs. The inner bark is rubbed and beaten down thoroughly with a smooth brass block to break up and homogenize the tissues and free the bark from the twigs. The peeler slits the bark with a small curved knife called a *kokaththa* and frees it in one piece. The pieces of bark are rolled in layers, telescope style, to produce long "quills" that are then left to dry for several days. Then the cinnamon quills, which have acquired a crackling, papery texture and the true cinnamon color, are trimmed to precisely forty-two inches.

In Sri Lanka, cinnamon peeling is still a specialized occupation that belongs to the Salagama caste, who carry on the traditional practices. Their contact with the potency of cinnamon's aroma is captured exquisitely in Michael Ondaatje's poem "The Cinnamon Peeler":

> *If I were a cinnamon peeler*
> *I would ride your bed*
> *and leave the yellow bark dust*
> *on your pillow.*
>
> *Your breasts and shoulders would reek*
> *you could never walk through markets*
> *without the profession of my fingers*
> *floating over you. The blind would*

stumble certain of whom they approached
though you might bathe
under rain gutters, monsoon.

. . .

You touched
your belly to my hands
in the dry air and said
I am the cinnamon
peeler's wife. Smell me.

Ondaatje imagines the powerful odor of the spice, freed from the bark, concentrated on the cinnamon peeler's hands. Practically, if less poetically, the aromas of cinnamon and other spices can be extracted for use in food and perfume by steam distillation, which yields an essential oil that concentrates the aroma of the simple ground spice by a hundredfold or more. This concentration of aroma makes the essential oil especially appropriate for perfume use; it also renders it vastly more expensive than the spice, which is far easier to process, handle, package, and ship in large quantities—and naturally loses much of its concentration of essential oil to evaporation by the time it reaches the consumer.

I love knowing about the varied history of cinnamon, and I love sampling the infinite variations of its scent, from region to region, tree to tree, branch to branch, essence to essence, to find the one that is most beautiful to me. As I source any given essence, I rely ultimately on my sense of smell, not on cost, producer's descriptions, or any other extraneous information. As I've mentioned, inferior cinnamon and the inexpensive cassia oil extracted from the Chinese tree and often used in cinnamon flavoring have the one-note heat of Red Hots. For a while I used a very warm, sweet cinnamon oil from

Ceylon in my collection of chef's essences for cooking, but even though it was true cinnamon, it was a little too sharp for my taste. I hunted through different suppliers until I came across a carbon dioxide extraction of Javanese cinnamon that had all the complexity of the essential oil I'd been using but was softer and rounder; I felt it would blend better with other aromas and flavors.

I find a similar range of quality in other essential oils. For a long time, for example, I could find only ginger essences distilled from dried ginger root. Their aromas were warm, but even at their best they tended to be somewhat musty, with a darkness and heaviness to their taste and smell. I stumbled upon a grower in Jakarta who was distilling ginger from the fresh root, and this was a revelation, light and citrusy-bright, almost sweet. The fresh ginger oil was so different from the dried as to seem derived from a totally different spice. Black pepper essential oil was another revelation, an opportunity to experience all the flavor and aroma of black pepper with none of its distracting heat. Again I needed to test different varieties before settling on one from Madagascar, which has a quietly floral backnote that marries beautifully with the dry and slightly woody pepper aroma.

Like wine, spices and the natural essences derived from them have terroir—the distinctive nuances of the soil, air, and water where they are grown. In their case, it is usually an especially exotic terroir. But (as with wine) a particular source terroir is not a guarantee of quality. The quality also depends on the variety of the plant, its health and growing conditions, the harvesting process, how long the plant sits before it is distilled, and under what conditions and methods it is distilled. These days I buy my rose oil from a small grower in Turkey, and each time I run low, I hold my breath awaiting resupply, praying that my purveyor is still in business and still extracting rose essence as gorgeous as the last batch.

COMMON CURRENCY

From the beginning, aromatics were central to all important aspects of life: sustenance (as seasoning), beauty (as perfume), health (as medicine), and spiritual life (via their role in ritual). Many of the same materials that were used in cooking were also prized for making perfume, and the appetite for scent contributed significantly to the enduring commercial vibrancy of the Spice Route. The word "spice" derives from the Latin *species*, meaning a commodity of special distinction as compared with the ordinary articles of commerce. Spices, which we now tend to think of only for their culinary uses, then permeated almost the whole range of remedies and love charms. To the classical mind, they included ingredients in ointments, perfumed powders, cosmetics, incense, and drugs. Hippocrates prescribed pepper not for use in seasoning food but as a treatment for disease.

Spices, in their role as a major ingredient in incense, were an early forerunner of aromatherapy. Churches and homes were permeated with the odors of resinous spices, which contributed to a feeling of spiritual and emotional well-being. Whether the spices were consumed or inhaled—and whether for health, savor, adornment, or worship—they created an atmosphere of refinement. Universally needed and desired, they became universally valuable and, before long, symbolic of wealth and status.

As with the Dutch colonizers' later restrictions on cinnamon propagation and production, in the ancient world a complex system of classification and regulation grew up around the use of spices and the commerce in them, an index of their social value and importance. The Greek philosopher Theophrastus offered a system of categorizing spices in his treatise "On Odors," and some principles for blending them: "Spices must be mixed and fall into two categories: wet and dry, or liquids and solids. There were thus three combinations: solid with liquid, liquid with liquid, and solid with solid. There were also the third and fourth components, namely the fixative of both colour and scent such as orris root and saffron. . . . The object of spicing was to produce a pleasanter taste, as in wines, or a pleasanter smell, as in perfumes. Spices were used in the making of all unguents to thicken the oil and to add fragrance."

A more specialized use of spices was in the preparation of antidotes, especially for those in power, who were in frequent danger of being poisoned. An antidote prepared in 80 B.C. for King Mithridates VI, a formidable enemy of Rome, by his court physician, Crateuas, was said to comprise almost all the known spices of the day, including costus, iris, cardamom, anise, nard, cassia, silphium, styrax, castor, frankincense, myrrh, cinnamon leaves and bark, galbanum, saffron, and ginger. "When these were ground together, bound with honey and mixed with wine, they provided an antidote against the strongest poison," noted the Roman encyclopedist Celsus in his *De Medicina*, a compendium on medicine from about A.D. 20. The mixture was reputedly so effective that when Mithridates tried to poison himself to avoid being captured by the Roman general Pompey, he was unable to do so.

The Romans classified spices according to their preponderant flavors: sweet, hot, bitter, or astringent. Myrrh was considered to be bitter and astringent, cinnamon and costus hot and pungent—

though cassia was still hotter and more pungent, and astringent as well. Cardamom, calamus, and spikenard were hot. Spices were often steeped in sweet wine before they were used, for the mutual benefit of the wine and the spice.

A set of rules for running a perfume shop in Constantinople circa 900 is an indication of both the range of goods then on offer and the degree to which a full-fledged industry and its attendant need for regulation had been established.

REGULATIONS FOR PERFUMERS IN CONSTANTINOPLE

Every perfumer shall have his own shop, and not invade another's.

Members of the guild are to keep watch on one another to prevent the sale of adulterated products.

They are not to stock poor quality goods in their shops: a sweet smell and a bad smell do not go together.

They are to sell pepper, spikenard, cinnamon, aloe wood, ambergris, musk, frankincense, myrrh, balsam, indigo, dyers' herbs, lapis lazuli, fustic, storax, and in short any article used for perfumery and dyeing.

Their stalls shall be placed in a row between the Milestone and the revered icon of Christ that stands above the Bronze Arcade, so that the aroma may waft upward to the icon and at the same time fill the vestibule of the Royal Palace.

The most telling of the regulations is the last. Above all, the spiritual dimension of the commerce in spices was too important to be left to the mere whim of the merchants.

By the Middle Ages, spice merchants and apothecaries sold an ever-growing array of articles. The medieval European diet of the upper classes included more spices than any since, but their high prices were also driven by competition from medical, religious, and perfumery uses, and their mode of distribution was correspondingly complex. In fact, our modern term "grocer" originally meant a spice merchant who handled larger wholesale, or "gross," quantities. The image of the mortar and pestle remains a symbol of the druggist, reminding us that the pounding and grinding of aromatics to release and capture essential oils that could be blended together was a process common to the apothecary and the cook. "The medieval spice merchant or apothecary seems to have handled several kinds of products whose relation to each other is not all that clear: edible spices, medicine, sweets (including medicinal preparations but also candied fruit, sugar-coated nuts and spices, nougats, confectionary of all kinds), cordials (spiced and fortified wines), wax (candles and sealing wax), paper, and ink," notes Paul Freedman in *Out of the East*. "Such establishments might even sell pasta or gunpowder."

Often the best record we have of a merchant's stock-in-trade comes from the property inventories cataloged upon his death, to settle his estate. The detailed account book kept around 1378 by a Barcelona merchant covered more than two hundred different products and at least a hundred aromatics confected in diverse forms for various uses. "Postmortem surveys of London grocers' shops from the reign of Richard II (1377–1399) show that besides spices and drugs they might sell soap, honey, alum, lamp oil, seeds, pitch, and tar. . . . Particularly conspicuous are sugared luxuries, such as glazed or candied quince, anise, almonds, ginger, and even small birds (larks, for example)," Freedman reports. The stock left by a pharmacist from Dijon in 1439, he adds, included two dozen kinds of spices and "exotic items on the order of pearls, coral, aloe wood, camphor, ambergris, and frankincense."

For culinary purposes, along with cinnamon, which remained ever popular as the routes of trade expanded, black pepper, ginger, and saffron were the spices most in demand. From medieval times onward, no banquet was complete without dishes highly seasoned with these and other spices. In European cookbooks from the thirteenth to the fifteenth centuries, spices appear in 75 percent of the recipes. They were, however, beyond the purse of most people. An English account written in 1418 reveals that half a kilo of ginger cost as much as a sheep. Ginger, the pungent underground stem or rhizome of the plan *Zingiber officinale*, is native to India and Southeast Asia. The spice was introduced to Europe in Roman times and became popular as a food flavoring, a position of favor it still held in Elizabethan times—the queen herself is said to deserve credit for the invention of that all-time favorite the gingerbread man when she ordered small ginger-flavored cakes to be made in the shape of her noblemen.

Pepper, the small, berrylike fruit of the climbing pepper vine native to southern India, is still probably the most widely used spice in the world. In medieval England it functioned not only as seasoning but as currency: peppercorns, counted out one by one like coins, were used to pay rent, taxes, tolls, and wedding dowries. In 1439 a pound of cinnamon was recorded as equivalent to three days of a skilled laborer's wages—a deflation since Pliny's day, perhaps, but still a pretty penny—while a pound of saffron would have paid those wages for a month.

All this burgeoning value brought with it the usual pitfalls. The problem of fraud in the retailing of spices was immense. Quality was hard to control, particularly because spices, though durable, were not immortal, and unscrupulous merchants often passed off stale merchandise as fresh. A larger problem was adulteration—it was so much easier to pass off a counterfeit in the form of dust in a spice blend than in, say, jewels. Along with ground coconut shells, spices were commonly cut (or replaced) with all manner of appalling impostors, as Gibbs elaborated a century ago:

> Allspice, ground with burned shells and crackers, clove stems, charcoal, and mineral color
>
> Cayenne, cut with rice flour, stale ship stuff (low-grade wheat flour), yellow cornmeal, turmeric, and mineral red
>
> Cinnamon, adulterated with cassia, peas, starch, mustard hulls, turmeric, mineral-cracker dust, burned shells, charcoal
>
> Pepper, replaced with ground crackers, coconut shells, cayenne, peas, beans, yellow cornmeal, and refuse of all kinds

He implored his readers to do their part to improve this deplorable situation by being willing to pay accordingly for quality merchandise:

> To the consumer of spices, this should be said: Be willing your grocer should live and obtain a profit for his work. Do not compel him to handle adulterated goods by quoting him the price of his neighbor dealer who sells the adulterated stock. Spices of high order are more costly, but are cheap to the consumer by reason of excess of flavor and strength. Let your dealer know you can appreciate a good article and, if he handles adulterated goods, remind him "that he may fool all the people some of the time, and some of the people all of the time, but he can't fool all of the people all of the time."

Truly, the exorbitant cost of spices only added to their allure, for individuals and for nations, for scrupulous and unscrupulous alike. Foreign trade was fundamentally spice trade, as strong an economic driver in their time as oil is in our own. They were also the most highly prized of all luxury goods. The rich used them to excess and the rest in moderation at best. They were bequeathed together with other heirlooms, instant indicators of social status. As articles of luxury, comparable with gems and silk, they could be purchased only with gold, itself a precious commodity.

THE STUFF OF LEGEND

From the beginning, spices carried with them the aura of something beyond themselves; their singular, exotic, and gorgeous aromas touched a longing deeply rooted in the human psyche. It's this capacity, at once less tangible and more personal than the exceptional quality and social status they represented, that so dramatically increased their value, as Freedman observes:

> Not all consumers would have worked out in detail these overtones of meaning, but that is what makes successful consumer products so durable and powerful: the sense they convey of promise, pleasure, and virtue beyond their inevitably mundane uses. Spices were simultaneously valuable commodities, social signifiers of discriminating taste, pleasurable substances, and yet vessels of higher, even sacred meaning. They conveyed this through a fragrance sometimes sweet, sometimes sharp, sometimes rich, sometimes impossible to describe, but always delightful.

Spices were both scarce and exotic, defining characteristics of luxury. The fact that maps were hard to come by in premodern times as well and navigation an art known only to a select few heightened the aura of mystery and rarity. Even after the great spice routes had been developed, most people had only a fanciful idea where exactly the redolent substances they held in their hands had originated. The arrival of baskets of frankincense and myrrh resin, along with reddish-orange sticks of cinnamon, must have seemed as providential and precarious as the weather.

The thrill of acquiring such materials was also heightened by the massive effort required to obtain them. Yet impressive as were the actual lengths to which people went to procure aromatics, the epic tales of their exploits, like all epics, embroidered reality—to wit, Herodotus's fanciful account about collecting cinnamon from giant birds' nests. He likewise reported that frankincense was guarded by snakes, while cassia was patrolled by dangerous batlike creatures. In the second century, Pausanias went Herodotus one better, noting that Arabian balsam was so seductive that the vipers that guarded the plants, having eaten their fill, had grown tame and no longer venomous. As late as the seventeenth century, the legend mill was still in full force, spinning tales of Indians setting fire to pepper trees in order to drive out the snakes that guarded them—which is supposedly why peppercorns are black and shriveled.

By drawing a scrim of myth around the source of a commodity for which there was high demand, such tales may have served a practical function, providing a kind of camouflage for information about its origins. Surely there were simpler ways of keeping the competition at bay, though. Why, then, the need to exaggerate the mystique of these already exotic substances? Weren't the actual obstacles overcome in finding them dramatic enough?

The Greeks had already discovered a truth that seems very modern: As the exotic becomes familiar, our sense of adventure requires a farther horizon—some imagined place that is still unknown and mysterious and perhaps dangerous to us. So fables are necessary to keep pushing the edge of the known world beyond view. Aromatic ingredients carry at once the faraway, unfamiliar atmosphere of the places they come from and the aura of some mythical, magical place beyond, thus keeping the appetite for—and the promise of—the exotic alive, without ever quite satisfying it.

The stories about how aromatic spices were procured became part of what they were, and their attendant costliness became part of the lore as well—a lore that merchants, then as now, knew how to exploit. "Retailers often emphasize or exaggerate the difficulty of obtaining exotic (or putatively exotic) or high-quality ingredients—'rare' botanicals for perfumes, 'hand selected' or 'belting' leather for car seats, 'diver' scallops," comments Freedman. "When it isn't really known where a valued commodity comes from, this mystification is all the more plausible, tempting, and attractive."

Over time the lore acquired a life of its own. The names of the perfume ingredients and the places they came from—Tasmanian boronia, Bulgarian rose, Moroccan orange flower—took on the potency of the substances themselves, as if steeped in them. The power of scent to conjure the faraway became embedded in the language: we could catch a whiff of adventure in it, the tang of the exotic. Poets knew this well—Baudelaire wrote of "a perfume of green tamarind trees" that "melds in my soul with the sailors' song."

The rhetoric of exoticism persisted long past the reality. With the advent of mass cultivation and production of spices, they ceased to be luxury items. And with the introduction of synthetic scent—coumarin, smelling of new-mown hay, was commercially synthesized in 1868, followed by vanillin (synthetic vanilla)—perfume was transformed into a mass-produced commodity. Whatever initial caution the perfume community had about employing the new synthetics, which were not of the same quality as the naturals, it was overcome by their cheapness, their uniformity, and the freedom they allowed to approximate aromas that couldn't be harvested from natural materials—or that didn't exist in nature at all. Before long the synthetics had come to dominate the perfumer's palette, altering the character of fragrance blends irrevocably. Synthetic perfumes were long-lasting on the skin, their component notes ever present rather than evolving; they were cleaner and fresher and less layered, less earthy. And yet the powerfully emotive narratives that derived from the history of natural aromatic materials continued to be evoked by the manufacturers and promoters of the mass-produced, synthetic perfumes—poetic images of roses plucked at dawn (for entirely synthetic perfumes) or magical-sounding ingredients like black orchid (which cannot be scent-harvested).

This process was mirrored to a degree in the world of food, where the introduction of synthetic flavorings quickly developed into an increasing reliance on them and the products they made possible: cheap, uniform, mass-produced, processed foods with endless shelf life. It had its parallels in other consumer goods, too, such as fabric. And as with perfume, the romance of the rhetoric outlived the reality, from "French vanilla" ice cream to the exotic-sounding Qiana, the quintessential synthetic fabric, or the entirely made-up and luxe-sounding fragrance ingredient cashmerian. The enduring appeal of the language of the exotic is testament to our enduring hunger for it, no matter how corroded the linkage between the copywriter's verbiage and the substance itself. We continued to fall for that French vanilla ice cream or the perfume made from ersatz "Italian bergamot" even if it originated in a factory in New Jersey.

Of course, the high-quality originals still beckoned, a luxury for those who could afford to travel. You went to Paris to purchase French perfume or to see and shop from the fashion collections. Back home you tried on the gown or opened the gorgeous flask and remembered not only the perfume but the holiday—and the journey. The lengths to which you'd gone to obtain these items, the experience of acquiring them, was part of what made them luxurious.

But as travel and transportation became cheaper and goods and materials for which there was a demand began to be produced and imported from wherever it was cheapest to do so, less and less luster inhered to the process of bringing back things from elsewhere. The world grew knitted together to an almost stifling closeness; the faraway, the Other, seemed barely to exist. Shopping at a duty-free shop in an airport was not like finding a one-of-a-kind boutique on the Left Bank. These days you don't even have to leave home. You can shop the entire world from your iPad, thanks to Google. But, unfortunately, the world itself seems dismayingly more uniform, more prepackaged, more commodified, more watered down.

And yet: This appetite for adventure is intrinsic to our species, and it demands to be satisfied. "It is fantastic to work beside Yves," Loulou de La Falaise told *Vogue* after twenty years of collaboration with the consummate tastemaker Yves Saint Laurent. "We both believe fantasy is such a vital element of fashion. We tend to think of ourselves as gypsies who have just returned with a marvelous caravan of incredible finds from the exotic reaches of the earth. But we have to make the caravan ourselves. Our Orient is our imagination." We still long for authentic luxury, and there are still places to find it and ways to incorporate it into our lives.

LUXURY: THE CONQUEST OF THE SUPERFLUOUS

The conquest of the superfluous gives us a greater spiritual excitement than the conquest of the necessary. Man is a creation of desire, not a creation of need.

—GASTON BACHELARD, *The Psychoanalysis of Fire*

Luxuries are by definition not necessities, yet, as creatures of desire, we have "needed" them through all times and cultures. The grand reach of the spice trade was all the more remarkable for the fact that the goods it went to the ends of the earth to retrieve were in no strict sense essential. As John Keay comments in *The Spice Route*: "No vital industry depended on spices, and with the exception of medicine, no branch of the arts or sciences could not have managed without them." But he goes on to say, "That, though, is the point. In ages past, when utility was paramount, the allure of spices lay precisely in their glorious irrelevance. Rare enough to imply distinction and distinctive enough to be unmistakable, spices unashamedly announced themselves as luxuries. . . . To the nose and palate, they acted as silk to the touch, or as music to the ear."

Many of the luxuries we prize most highly today gratify the sense of smell, alone or in combination with our other senses: food, wine, leather goods, flowers. Our most purely aromatic luxury, perfume, serves no practical use at all; it is about pleasure and indulgence alone. Luxury is a conscious indulgence in sensuality, refinement, and grace, a sphere we enter when we set the practical necessities of life aside. Sometimes these beautiful sensual experiences are plentiful and inexpensive, and other times they are rare and costly. The taste of tomatoes and peaches in summertime, when they are at their lushest, ripest best, is a luxury. Does their being abundant and therefore cheap diminish the experience of a single glorious bite? As Coco Chanel said, "The opposite of luxury is not poverty, because in the houses of the poor you can smell a good pot-au-feu. The opposite is not simplicity, for there is beauty in the corn-stall and barn, often great simplicity in luxury. But there is nothing in vulgarity, its complete opposite." The preoccupation with status is one expression of this vulgarity. Status and luxury may coexist in the same object or not, but they are two different things. Luxury is the internal experience evoked by the way a material thing plays upon our senses; status is what that object signifies to others.

What goes in and out of style is status, not luxury per se. Status is akin to a code in a given culture, agreed upon by the powerful and subject to change. Once spices became widely available, they were no longer seen as luxury items, or perhaps it is more accurate to say that they no longer garnered substantial status. If they therefore seemed less luxurious, it was not because they had lost sensual potency but because people had allowed their attachment to the external perception of status to eclipse their internal experience of sensuality. In such an eclipse, the experiential value of the thing itself plummets along with its status; as it becomes a mass-market item, people lose their reverence for it, fueling demand only for the imitation or the low-quality version of it. Thus, as Timothy Morton observes of cinnamon in *The Poetics of Spice*, "Yesterday's banquet ingredient becomes today's Dunkin' Donuts apple cinnamon item."

The titan of the spice trade is reduced to a mere flavoring, and a synthetic flavoring at that.

Yet the hunger for authenticity, while it goes dormant from time to time, in this place or that, does not die. In the world of food, decline—the near starvation of the taste buds—has given rise to a new beginning, a revived interest in (and value attached to) food that has now waxed long and strong enough to make it clear that this is no fad but a culinary revolution. It's true that our connection to what we eat was never as degraded as our connection to what we smell. After all, we had to survive on the stuff, and man cannot live by chemicals alone (though God knows we've tried). The rise of consciousness about food has unquestionably restored a deeper awareness of the connection between provenance and product. More people now know that the best vanillas come from Tahiti, great olive oil from Tuscany, fine chocolate from Madagascar, and incomparable goat cheese from France. And when you look for cinnamon on your grocery store's shelves, you may even find the better-quality Saigon cinnamon, which has a higher concentration of essential oil. If you shop in a particularly upscale grocery store, you might even encounter it in the rough reddish-brown scrolls in which it played its part as one of the ancient heroes of the spice trade.

I believe we are on the cusp of a revolution with fragrance as well—and I am thrilled to help lead the charge. I think we can borrow a leaf from the locavore and Slow Food movements, through which many people have developed an awareness of the history and provenance of the ingredients that go into their food, with a concomitant awakening of their concern for quality: heirloom tomatoes rather than mass-produced greenhouse varieties that never seem to ripen, artisanal goat cheese rather than Cheez Whiz. The important thing about the materials we consume—edible, olfactory, and otherwise—isn't that they come from here, there, or any particular producer or terroir, which, like exoticism, can be just a savvy marketing story that describes something of value—not necessarily the product at hand. I think of terroir this way: as the unique qualities that are lost when we try to mass-produce goods, sacrificing nuance for uniformity and losing the intimacy that becomes part of an artisanal product as it is created and passed to the consumer. The experience of luxury is heightened by an awareness of the stories behind the materials and the creation process itself.

The substances that make up the world I live and work in really do still come from particular places, often in hidden parts of the world. So although I rarely travel, the exotic is not just in my imagination. Even now I use the fresh ginger essence from Indonesia, the linden blossom from Bulgaria, the rose from Turkey. I love holding in my hand materials that have been used since ancient times, in every culture across the planet. Their intense intrinsic beauty has sustained their preciousness through the ages. And smelling them is to smell all the stories ever told about how they were discovered, why they were valued, where they came from, and what they were used for. Luxury starts with superior materials, but their stories exalt them, in a kind of alchemy that transforms base into precious.

Each time I work with these aromatic essences, I experience anew the exquisite rush of touching the unusual and the extraordinary, knowing that this material is marked by the locale from which it comes, and also by its rich history. It is an absolute thrill to feel a connection, through the use of aromatic materials, all the way back to biblical times—to the perfume formula that God gave Moses on top of Mount Sinai, for example: "Take thou also unto thee principal spices of pure myrrh five hundred shekels, and of sweet cinnamon half so much, even two hundred and fifty shekels, and of sweet calamus two hundred and fifty shekels, and of cassia five hundred shekels. . . . And thou shalt make it a perfume, a confection after the art of the apothecary, tempered together, pure and holy" (Exodus 30:23–35). Many of the materials that call to us today are the same as those that Cleopatra ran her hands through, that were used to wash the feet of Jesus, or to ward off the Plague. These materials are a direct link to the past. Their stories transport us far away in time and place—and how we yearn to go!

Above all we can re-inhabit an attitude of times gone by, seeing aromatic ingredients as not only health-promoting but delicious and moving and beautiful: luxurious. Fragrance straddles the line between goods and experiences. Experiences are memorable and live on, often for a long time, inside those who have them. Wearing perfume is one of the original luxury experiences, an experience that is still about how you feel when you're wearing the perfume, your senses deeply engaged in the nuances of this particular aroma, at this very moment. No matter how thoroughly explored the world is, "quality" is only ever specific. The language of desire is that of specific things, not groups of things. It is about this antique rose-cut diamond ring and not that one, or this fresh ginger oil and not that one. Luxury incites excitement and gratitude for the deep thrill and sense of specialness it grants us when we are brushed by it. When we taste it. When we inhale it.

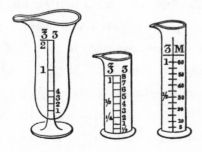

As I mentioned, knowing a little about how scent works will dramatically increase your awareness and appreciation of it, making you an alert and sophisticated "nose." And there's no better place to begin than by familiarizing yourself with the nuances of the aromatic materials themselves.

THE CINNAMON VARIATIONS

Cinnamon is the Julia Roberts of spices: that blinding smile makes it hard to take in the rest of this dazzlingly beautiful and talented woman. Cinnamon's warm, immediate beauty can eclipse its subtler aspects. If you smell it with a new nose, an attentive nose, you'll discover a miracle hiding in plain sight.

There is no better way to understand a given aroma than to smell many different versions of it. It's as if you were metaphorically walking in a circle around the aroma, seeing it from every side and perceiving thrilling subtleties of difference, each of them coloring your assessment of its overall character. Smelling real cinnamon is a particularly easy way to appreciate the unique layering of smells that make fine cinnamon such a prized aroma and taste.

Get samples of cinnamon and cassia in a number of different forms: powdered cassia, stick cassia, powdered cinnamon, stick cinnamon. Order very small quantities of essential oils from different purveyors, harvested from different countries, at different prices. (See Sources; many online suppliers sell tiny sample sizes of oils.) Smell each in turn. With the essential oils, use scent strips, lengths of stiff, unscented, absorbent white paper around five inches long and a half-inch wide. (You may have seen them at perfume counters in department stores.) Place a drop of each essential oil on its own scent strip, label the strips and return to smell them periodically over a couple of hours. Don't let yourself dwell on clichéd associations like doughnuts and sweet rolls. Cinnamon's marriage with sugar and doughy pastries is a product of its immediate, encompassing warmth. Like Julia Roberts's ravishing smile, it tends to eclipse other aspects of cinnamon's ravishing beauty. Try to separate cinnamon from its role in food and appreciate it for its own complex and marvelous smell. Do run your mind over its history, though, and its lore—they keep the magic of scent alive for us. Let these stories infuse your encounter with cinnamon as its warm, rich spiciness thrills you anew.

An 1873 guide lists the following cosmetic and health uses for common aromatics:

Ginger lozenges—If strong, are sometimes serviceable in slight toothache; also useful in flatulence, defective appetite and dyspepsia, &c.

Myrrh lozenges—Used in tender, spongy, and ulcerated gums; also to fix loose teeth.

Opium lozenges—Sucked as an anodyne and hypnotic, in toothache, face ache, &c.; also to allay tickling and irritation.

Soda and peppermint lozenges—These are antacid and carminative, and, when really good, are also often serviceable in flatulence, nausea, and toothache.

Lozenges to sweeten the breath—Lavender, musk, orange flower, Orris root, rose, and violet.

Lozenges for smokers—Cinnamon, cloves, orange, peppermint, and vanilla.

Lozenges for tipplers to cover the fumes of liquor—Aniseed, caraway, lemon, peppermint.

Jot down your impressions of each form you smell—no more than a few words that capture your experience of it. Compound words that you come up with yourself are often the most accurate and the most meaningful descriptors—for example, *dusty-woody, warm-spicy, sweetly tenacious, dry-powdery, uniformly candy-sweet,* and so on. You will quickly become a cinnamon connoisseur. More important, you will be experiencing scent with a sense of curiosity and wonder—even a scent as familiar and common as cinnamon.

SCENT REGISTERS

In addition to being categorized by their forms—essential oils, concretes, absolutes, and CO_2 extractions—essences for perfume and flavor are classified according to their relative volatility, or how long they remain perceptibly fragrant before the scent fades. Top (or head) notes are the most fleeting; less so are middle (or heart) notes; and base notes last longest. This volatility is the most important quality to consider when creating a fragrance or a flavor. Top notes will strike your senses first, followed by middle notes and then base notes.

> **TOP NOTES** are bright and fleeting and often familiar to
> us from cooking. They include herbs and spices, such as
> coriander, spearmint, ginger, black pepper, cardamom,
> juniper, and tarragon; and citruses, such as lime, bitter
> orange, blood orange, tangerine, and pink grapefruit.
> Most essential oils are top notes.
> **MIDDLE NOTES** give body to blends, imparting warmth
> and fullness. They include absolutes and essential oils.
> Many are floral absolutes, such as rose, jasmine,

lavender, orange flower, and tuberose; and floral essential oils, such as geranium and ylang-ylang. Cinnamon lives here, along with many other spices, such as pimiento berry (also called allspice) and nutmeg.

BASE NOTES, intense and profound, are often thick and syrupy. Most are derived from bark (sandalwood), roots (angelica), resins (frankincense and myrrh), lichens (oakmoss), saps (benzoin and Peru balsam), and grasses (patchouli and vetiver). The animal essences—ambergris, civet, musk, and so on—reside here, too.

You can determine for yourself whether an essence is a top, a middle, or a base note. Write the name of the essence you are sampling on one end of a scent strip and dip the other into the essence in question. Over the next several hours, smell the strip intermittently, first at intervals of no more than ten minutes, then less frequently. Top notes lose most of their scent within about thirty minutes, middle notes over the course of a couple of hours, and strong base notes take many hours to evolve fully, sometimes lasting for days.

If you have decided to acquire and experiment with natural essences, whether for scent or flavor or both, it is helpful to group them by register, so that you can visualize blending possibilities as you build a fragrance or flavor. In the perfume "organ"—the narrow, stepped shelves where I store my extensive collection of ingredients—I keep each register in its own section and color-code the labels of each bottle and cap accordingly.

CINNAMON AND SPICES IN COMPOSITION

Cinnamon has higher odor intensity than its close compatriots in the spice world—ginger, black pepper, allspice, nutmeg. Cinnamon opens with warm, sweet, candylike notes, and finishes with a powdery, dry wood. Allspice has a tealike undertone and smells like a blend of clove, nutmeg, and cinnamon. Fresh ginger is a revelation of lemon and citrus notes growing warmer and spicier, with a dry backnote. Black pepper essential oil has all the woody, spicy aroma of the actual peppercorn and none of the heat, plus a slightly floral background note that is imperceptible in the spice itself. And warm nutmeg is suave, with bittersweet balsamic backnotes.

The register of the spice oils usually changes, along with the texture and shape of the aroma, depending on which version of the oil you are using—an essential oil, an absolute, or a CO_2 extraction. Although there are inconsistencies, most of their essential oil and CO_2 versions are top notes, whereas as absolutes they tend to be middle or base notes. In some forms some of the spices have a sharpness that I feel detracts from their aromatic beauty. For both flavor and scent, therefore, I prefer the CO_2 extractions of cinnamon and allspice and the absolute of nutmeg, which have greater warmth, softness, and roundness to them. But in the case of fresh ginger and black pepper, which are top notes, a pungent sharpness is just what is needed, and for that reason I like to use them as essential oils.

The spice essences work beautifully as modifiers for floral notes and give definition to resins and softer notes. Their individual characters must be respected, however. I like to use black pepper when I want to sharpen a blend without adding a very pronounced note. Fresh ginger and allspice make their presence known with slightly greater intensity, but they generally ingratiate themselves easily in most perfumes. Pronounced aromas like cinnamon, cardamom, and clove, however, will dominate a fragrance unless dosed very lightly. And when carelessly mixed together, spice essences will lose their individual identities and devolve into a clichéd potpourri.

A nineteenth-century recipe for smelling salts:

Sesquicarbonate of ammonia commonly passes under this name with the vulgar, and, with the addition of a few drops of essential oil, is frequently employed to fill "smelling-bottles." Its pungency, however, is neither so great nor so durable as that of the true and neutral "carbonate of ammonia." The latter salt continues to change in composition, and preserves its pungency as long as a particle of it remains unvolatilized. The portion only which flies off suffers decomposition, as it volatilizes, separating into "gaseous ammonia" and "carbonic acid."
Take of

carbonate of ammonia (crushed small) 1 pound
oil of lavender 1 ounce
oil of bergamot 1 ounce
oil of cloves 2 fluid drachms
oil of cassia 1 fluid drachm

Rub them thoroughly together, sublime at a very gentle heat, into a well cooled receiver, and at once put the product into a well-stoppered bottle or bottles. The sublimation may be omitted, but the quality of the product suffers.

THE ZEN OF PERFUME MAKING

For readers who want to try their hand at blending, at the end of each chapter I've provided a recipe for a perfume that builds on the rock-star scent we've been following. But before we embark on making perfume, I want to introduce you to my basic approach to the craft of blending. Having an orderly system in place is especially important, because the narcotic sensuality of smelling natural essences can make it hard to keep track of what you're doing; in fact, the interplay of order and a dream-like, instinctive spontaneity is the paradox that beats at the heart of the process of creating beautiful perfume. Here, then, are the principles and practices—acquired through much trial and error!—that I follow every time I make a perfume:

- **The whole process should be beautiful.** The process of creating perfume should mirror the end result. You honor the materials and the process by your attention to beauty when choosing and labeling the bottles in which you store your essences and the tools for mixing and measuring them. Your tools need not be complicated or expensive, but they should be chosen with care.
- **Adhere to a strict "budget."** Every perfume is a blend of base, middle, and top notes. When blending perfumes in alcohol, which creates a different evolution, I like to adhere to a strict budget of drops for each register. For example, if I use a total of ten drops (a good round number) of essence to make up the base, then I also use ten drops of essence to make the middle and ten drops of essence in the top. (Solid perfumes and other oil-based perfumes evolve differently

and lend themselves to less intricate combinations and allow for greater flexibility in proportions.) The temptation to tamper with the number of drops is enormous, particularly if the blend is going well. A budget fosters creative discipline and focus, and it results in an architecturally sound perfume. Think of it like a financial budget: with only ten drops to spend, you want to make every drop worthwhile.

- **Write down everything you do while you're blending.** Keeping good notes allows you to retrace your steps, correct mistakes, and create variations. Never count on your memory. When you're in the throes of blending, you will be in an altered state of consciousness that is not easily recalled once you're out of it.

- **Experience the perfume on the skin after the addition of each new essence.** Natural fragrance blossoms on the body and should be smelled on your skin. The best way to smell a natural perfume is to put a drop or two on the back of your clean, unfragranced hand. Rub the perfume twice to make the alcohol disperse, so that you can smell the perfume itself. Close your eyes when you smell the perfume, and focus on what strikes you and in what order. Take note of the complexity, texture, and shape of the scent. When you're composing, you will gain important knowledge this way at every stage of composition: how the newest addition has altered the direction of the blend and whether or not you like where it's going. The last essence added always dominates the mixture for a few moments before it settles back down, but in those first few moments its louder volume gives you the opportunity to understand clearly what the new arrival brings to the blend. Smelling at every step also allows you to pinpoint where a blend has gone astray, so that you can remake it to the last good step and take a different path from there. I indicate this with the notation "GTH," which means "good to here." I find that using such a system allows me to take risks in composing, because I know I will be able to trace my way back to solid ground.

- **Plan on failing and redoing.** Don't count on concocting something good the first several times you make a blend. Plan on revising and allow yourself the freedom of failure. Creating a good perfume is a process.

- **Be real, let go.** From teaching perfumery I know how hard this one is. Time and time again, I see students have trouble admitting to themselves that a perfume doesn't work and letting it go. If a blend is bad, don't try to tell yourself it's beautiful. Dump it. You need to make lots of bad perfumes to make a good one. Be as honest as you possibly can about what you've made. The world does not need more bad perfumes.

- **Spend money to create something good.** Accept the idea that you will be spending lots of money on essences in the process of learning to make good perfume and that money is not wasted. Get really comfortable with that idea. It will give you the freedom to take risks and be honest with yourself.

- **Creating perfume mirrors life.** Budget, personalities, money, failure, honesty, and sensuality—although perfume is an almost intangible thing, it is a mirror of and a vehicle for the tensions and complexities of life.

CARRIERS: ALCOHOL, JOJOBA, AND FRACTIONATED COCONUT OIL

Blending essences to make perfume requires a carrier, or medium. By far the most common carrier for liquid perfumes is 190-proof undenatured ethyl alcohol (see Sources), which mixes completely with essential oils and absolutes and will dilute the thickest of resins, balsams, and concretes. It also helps to lift and diffuse the essences, allowing them to blossom as aromas. Alcohol is very flammable and should be stored well away from sunlight and heaters. Isopropyl (or rubbing) alcohol is strong-smelling and unsuitable for perfume making.

Jojoba oil (pronounced *ho-HO-ba*) is made from the seeds of a desert shrub and is a lovely golden color, with no fragrance of its own, and is much less prone to rancidity and oxidation than other oils. Jojoba oil is actually a wax, not a liquid oil, that closely resembles human sebum and is an excellent moisturizer. It is widely available in natural-food stores and online (see Sources).

Fractionated coconut oil is a light, fluid, nongreasy, nonstaining, liquid oil. It contains only the medium-chain triglycerides (MCTs) of coconut oil, produced through the hydrolysis of coconut oil, which is then fractionated by steam distillation to isolate the MCTs. It has an indefinite shelf life and is also widely available online (see Sources).

EQUIPMENT

BEAKERS Tiny beakers to use for blending are available from lab supply houses or on eBay (see Sources). Look for beakers wide enough that they won't easily tip over, and calibrated every 5 milliliters (ml). If you don't have metrically calibrated beakers, here are the liquid volume equivalents in common cooking measurements:

.6 milliliters = ⅛ teaspoon

1.25 milliliters = ¼ teaspoon

2.5 milliliters = ½ teaspoon

3.75 milliliters = ¾ teaspoon

5 milliliters = 1 teaspoon

15 milliliters = 1 tablespoon

FILTERS A simple plastic coffee filter and unbleached filter paper can be used for straining perfumes. Before straining another perfume, replace the filter paper and rinse the filter in clean alcohol to eliminate the previous scent.

GLASS DROPPERS Use glass droppers, not plastic pipettes, for measuring essences and other ingredients. They can be bought in a drugstore or by the dozen online (see Sources).

HOT PLATE You can use a gas or an electric burner for melting wax if you watch carefully to make sure the wax doesn't burn. If you become more serious about making solid perfume, get a small specialized hot plate from a laboratory supply company or on eBay.

NONMETALLIC PAN Use a nonmetallic pan for melting wax. Ceramic or glass is best. Lab supply houses sell tiny heatproof ceramic dishes that are perfect for small batches of solid perfume. Little ramekins or soufflé dishes from your kitchen are also suitable.

GETTING STARTED MAKING PERFUME

I like to keep the process of creating perfume as simple and hands-on as possible. I don't like distance from the essences and prefer to use them at full strength, smelling their pure, intense aromas and watching as they combine with the other essences in the alcohol or oil, drop by drop.

Fold a paper towel in half. Put your beaker or bowl or whatever receptacle you are going to blend in on the left side of it. On the right, place two clean, dry droppers and a shot glass filled with alcohol; this is for cleaning the droppers. Measure whatever carrier you are working with, such as jojoba oil or alcohol, and add it to the blending receptacle on the left. (If you're left-handed, reverse the placement if that seems more comfortable to you.)

Have all your materials out and available—including the essences you are intending to use—when creating a perfume. Establish a system of placing the bottles that allows you to know at a glance which essences you have already put into the blend and which are waiting to be put in. My system is to keep right in front of me the essences that have not gone into the blend and push to the opposite end of the table those that have already been added. When you are creating, don't answer the phone or allow other unnecessary interruptions; it is so easy to lose your place and put in a double dose of a crucial essence, or no dose at all.

Add the essences to the container with the fragrance medium a drop at a time, completing each register before you move to the next. Always start with the base notes, then move to the middle notes and finally the top notes. If you wait to compose the base notes till last, you will not be able to smell the top notes clearly or understand their effect on the blend.

When you finish with each essence, place the dropper that you've just used in the beaker of alcohol and pump it to clean it, leaving it in the beaker and filled with alcohol until next use. When you need it again, empty it and wipe it on the paper towel to make sure it is clean inside and out, so that you don't cross-contaminate your oils.

As you stop following recipes and begin to create fragrances on your own, a good way to start is to choose one top note, one middle note, and one base note that you are attracted to and want to work with. Think of this as a central chord around which you will build the perfume, three essences that have a relationship with one another. Smell them to try to figure out what that relationship is and how to develop it in each register. Choose two additional essences in each register to complement these central notes. At each level start by adding around eight of your ten budgeted drops in whatever proportion seems right to you, keeping in mind that lopsided quantities generally result in more interesting blends than even proportions. Smell the perfume after each addition, by putting a drop on a fresh part of your hand each time and rubbing it in. This way you will be able to experience the aromatic changes each essence brings to the blend and decide what to do next, adjusting your choices or the proportions of essences you've already used. With a couple of drops in reserve in your budget for each register, you will be able to adjust the blend once you see how the essences are interacting.

In order to figure out which essences to include, I often use scent strips or the caps of the bottles themselves. Smelling them in conjunction with the existing blend helps you figure out what to add next, the way smelling jars of spices helps you figure out which seasoning to add next to a soup or stew, and in what amounts. Top notes are always the hardest to get right, the same way you can screw up a recipe at the end by adding too much pepper or lemon juice. Remember, if that happens, you can always back up to a point when the blend was good and redo it—assuming you've remembered to keep careful notes!

Two imaginatively titled recipes from an 1896 *Techno-Chemical Receipt Book*:

Parfait d'Amour Essence: Dissolve ½ fluid ounce of oil of cinnamon, 6 fluid drachms each of oil of cardamon, oil of rosemary, and anise seed oil, and 20 minims each of oil of lemon, oil of orange, oil of cloves, oil of camomile, and oil of lavender in 1¼ quarts of rectified spirit of 90 per cent. Shake the solution thoroughly and filter it.

Elixir of Life: Dissolve 2 fluid ounces of oil of wormwood and 1 fluid ounce each of oil of cardamon, oil of calamus, oil of nutmeg, and oil of orange peel, in 3½ gallons of alcohol 90 per cent strong, and add ¾ of a gallon of water to the solution. Color the fluid brown with burned sugar.

*A*MBER SPICE PERFUME

In this chapter and those that follow, I'll introduce each perfume recipe first by describing the chord—one top, one middle, and one base note—at the heart of it. Then I'll walk you through recipes for a body oil, a solid perfume, and a simple oil-based perfume built around that chord. This basic structure with few ingredients works well for oil-based solid and liquid perfumes; those mediums are more forgiving than alcohol, which tends to heighten and expose the essences and therefore requires a more complicated design to ensure that the fragrance notes evolve gracefully on the skin over time. Once we have gotten the hang of the chord in an oil-based perfume, we'll elaborate on it in a complex, architecturally sound liquid perfume in alcohol, with fully developed top, middle, and base chords.

The simple chord we will be starting with is built upon an amber base note, combined with cinnamon and bois de rose. Cinnamon is straightforward; bois de rose and amber need some explaining. Bois de rose is a dry, rosy wood. Amber has nothing to do with the semiprecious fossilized resin of the same name. It is an ancient fragrance note that is built upon labdanum, the resinous exudation of the rockrose plant, often in combination with vanilla, to create a cozy, sweet, powdery, leathery aroma. Here the cinnamon gives it a spicy twist and the lime an uplifting splash of freshness in the opening of the fragrance.

SOLID PERFUME

    8 milliliters jojoba oil
    Heaping ½ teaspoon grated beeswax
    10 drops labdanum absolute
    2 drops cinnamon
    4 drops rose absolute (optional)
    10 drops lime

A solid perfume is magic: an olfactory, visual, and tactile pleasure all at once. The jojoba oil and beeswax give the perfume a lovely golden color, and the beeswax contributes a soft honey scent. The texture is important—too high a proportion of beeswax to jojoba oil results in a perfume that is hard and unyielding, too little wax and the perfume is runny. The ideal texture is like that of a good lipstick—firm but creamy and adhering to your finger easily. To put on a solid perfume, run the ball of your finger back and forth across the surface of the perfume—

don't use the tip of your finger to scoop it, or you'll leave an unsightly marred surface.

The best shape of a case for solid perfume should be round or oval. Flattish and shallow is better than deep. Small compacts and pillboxes are perfect, and you can find lovely ones at flea markets and on eBay. Make sure the case latches securely.

To make the solid perfume, measure the jojoba oil into a small beaker or bowl and add the essences. Melt the grated beeswax in a small pan or ramekin over a hot plate or stove. Quickly add the jojoba-and-essence mixture to the melted wax, stirring over the heat for around 10 seconds to thoroughly mix. Expose the perfume mixture to the heat for the shortest possible period of time. Pour into your perfume case and close immediately. Do not touch or move for 15 minutes while it cools.

OIL-BASED PERFUME

A simple oil-based perfume in fractionated coconut oil is lighter and much less greasy than a body oil, and the top notes have more lift and elegance as the composition evolves on the wearer's skin over time.

15 milliliters fractionated coconut oil
10 drops labdanum absolute
2 drops cinnamon
4 drops rose absolute (optional)
6 drops lime

Measure 15 milliliters of fractionated coconut oil into a beaker or small glass. Add the drops of labdanum, cinnamon, rose absolute, and lime, stirring after each addition. Put into a small bottle and seal tightly.

BODY OIL

Body oils are heavier than perfumes, more like a moisturizer that leaves a subtle veil of scent on the skin. The notes are flatter and weaker and harder to distinguish.

> 20 milliliters jojoba oil
> 10 milliliters fractionated coconut oil
> 10 drops labdanum absolute
> 10 drops rose absolute
> 2 drops cinnamon
> 10 drops bois de rose

Mix together the jojoba oil and fractionated coconut oil and then add the drops of labdanum absolute, rose, cinnamon, and bois de rose. Pour into a bottle and seal tightly. Shake to remix before each use.

ALCOHOL-BASED PERFUME

> 8 milliliters perfume alcohol

BASE NOTES

> 2 drops benzoin absolute
> 2 drops vanilla absolute
> 8 drops labdanum absolute

MIDDLE NOTES

> 7 drops rose absolute
> 2 drops cinnamon essential oil
> 1 drop jasmine absolute

2 drops fresh ginger essential oil

6 drops lime essential oil

3 drops bois de rose essential oil

The base of this perfume is a warm, ambery note built around labda-num, sweetened with vanilla and benzoin. Universally beloved, it is com-forting, sweet, balsamic, with notes of leather and vanilla. The center of the perfume is rose, spiced with cinnamon and supported with a touch of jasmine—like a luscious, spicy rose plucked from your back garden. Lime contributes juicy citrus notes and highlights all the other top notes. Bois de rose extends the rose notes to the opening of the perfume and, with the fresh ginger, creates a mouthwatering opening to the perfume.

Remember to smell after each step. When you have finished the blend, wait 5 minutes and smell it again. Notice how it has changed. Try again after 10 more minutes, then again after half an hour. This will give you a good reading of the evolution of the perfume on your skin.

CREATING FLAVOR

Creating a flavor is fun and simple—it requires only an essence or two—and the rewards are immediate. You can flavor with drops of essential oils or create a spray by diluting the oils in vodka. The sprays are best for adding to finished dishes and using in drinks—for example, to flavor tea, ice cream, sparkling wine, chocolate, vinegar, and olive oil. Drops of undiluted essential oils can create magic when added during the cook-ing process to baked goods, puddings, roasted meats, chicken, and fish.

The spice essences are particularly easy to use. Black pepper essen-tial oil is as universally welcome as is a pepper grinder on the dining room table—and for those who like food spicy but not hot, it gives you all the flavor with none of the heat (or specks of black!). Fresh ginger essential oil likewise is easy to use and a welcome addition to teas, vegetables, salads, desserts, puddings, and ice cream. Since black pepper and fresh ginger are top notes, they will be the first flavor you taste. Nutmeg, allspice, and cinnamon, being middle notes, come to prominence next, sliding in under a top note like lemon or orange.

Cinnamon is a particular pleasure to use in creating flavor, because behind that big, warm smile lies a wealth of nuance that enhances both sweet and savory. You might think of cinnamon as a reconciliation of opposites: sweet yet earthy, exotic yet comforting. It brings out the warmth in sweet foods like pears, oranges, figs, apples, and carrots. It rounds out salty and spicy tastes as well, for example when matched with cumin in lamb or beef dishes, or in a Moroccan carrot salad. Cinnamon's spicy sweetness pairs well with nutmeg, creating an allspice-like taste that you can bend toward one direction or the other, depending on the proportion of each. Cinnamon also pairs well with squash, apricots, coffee, chocolate, all meats and poultry, tomato, lemon, lime, and vanilla.

Try the following:

- Add a drop of cinnamon oil and a drop of orange oil to a pot of aromatic black tea to make your own freshly vibrant "Constant Comment" (a tea blend I associate with my college years).
- Add a drop of cinnamon oil to vanilla extract, then add a drop of that to a glass of champagne.
- Spray a cinnamon-essence/vodka blend over poached pears, chocolate, vanilla ice cream, baked apples, or roasted lamb.

• Add one drop each of cardamom, cinnamon, and allspice essential oils to a pot of black tea to make homemade chai.

"COCA-COLA"

The original formula for Coca-Cola featured essential oils of cinnamon and vanilla, along with orange, lime, lemon, and nutmeg. One reason that it was so addictive, literally, was that it also contained cocaine.

2 drops lime essential oil

2 drops orange essential oil

1 drop lemon essential oil

1 drop nutmeg essential oil

1 drop cinnamon essential oil

2 teaspoons vanilla extract

Drop the essential oils into the vanilla extract. Sweeten to taste with sugar, stevia, or honey, and add to 8 ounces of bubbly water.

I think of this 1891 recipe for "Sydenham's Laudanum" as an early attempt at Coca-Cola:

Opium, 2 ounces; saffron, 1 ounce; bruised cinnamon, bruised cloves, each one drachm; Sherry wine, 1 pint. Mix and macerate for 15 days and filter. 20 drops are equal to one grain of opium.

*A seventeenth-century engraving of flowering spearmint
by Dutch botanist Abraham Munting.*

CHAPTER THREE

THERE'S NO SMELL LIKE HOME

When we go to live in the house of memory, the real world vanishes all at once. What are the houses on our street worth compared to the house of our birth, that house of total interiority, which gave us our sense of inwardness? That house is remote, is lost, we no longer live in it, we are only too sure that we will never live in it again. And so it is more than a memory. It is a house of dreams.

—GASTON BACHELARD, "The Oneiric House"

int was so revered by the ancient Greeks for its delightful aroma and mildly stinging taste that they named the plant after Minthe, a beautiful river nymph who attracted the amorous attention of Hades, god of the underworld. When Hades' wife, Persephone, found out, she flew into a fit of rage and turned Minthe into a plant, so that people would trample her. Unable to undo the spell, Hades gave Minthe a wonderful aroma so that he could smell her when they trod on her.

If cinnamon represents the journey of a thousand miles to reach our doorstep, mint is an old familiar. Unlike the exotic spices that came from parts unknown, mint and other herbs were, since ancient times, native, domestic, homegrown the world over. As such they are entwined with—and evocative of—the many meanings of home.

DOMESTIC ARTS

In the early sixteenth century, a new kind of book appeared in Europe. These books, which came to be known as Books of Secrets, were popular compendiums that professed to divulge to the reader the secrets of nature, culled from ancient sources of knowledge and wisdom. Fabulous assemblages of formulas, advice, and information about a broad range of practical arts, they commingled what William Eamon describes in *Science and the Secrets of Nature* as "traditional lore concerning the occult properties of plants, stones, and animals, along with miscellaneous craft and medicinal recipes, alchemical formulas, and 'experiments' to produce marvelous effects through magic." In fact, much of their content was borrowed (and reborrowed) from ancient texts, but "they also included material based upon indigenous folk traditions, the accumulated experience of practitioners, and the discoveries of medieval 'experimenters.'"

The most famous Book of Secrets was the sixteenth-century *Secrets of Alexis of Piedmont*. Alexis of Piedmont, or Alessio Piemontese, was the pseudonym of an Italian physician and alchemist. More than seventy editions of his book were published in at least seven languages, and two centuries later the book was still being reprinted. It was made up of some 350 recipes—although they were not quite recipes in the contemporary sense. There was a chapter of formulas for honey-based concoctions and preserves, to be sure. But a third of the book was taken up with formulas that were medicinal—remedies for common ailments—rather than culinary. There was a chapter of recipes for perfumes and other scented ingredients—lotions, soaps, and body powders. And it wasn't just the culinary and perfume recipes that called for spices and natural essences. There was a mouthwash of benzoin, cinnamon, rosemary, and myrrh, for example.

These books' closest descendant in the modern world is the cookbook. But the very idea of a cookbook is a recent one. To the premodern mind, there was no clear distinction between food and medicine and craft, the domestic and healing and creative arts. Substances derived from plants and animals and minerals were "simples"—building blocks that could be combined to form different compounds. "Pharmacists had to know how to grind up mixtures of simples according to medical instructions or their own ingenuity," writes Paul Freedman in *Out of the East*. "Pounding and grinding together these aromatic products was a tedious task and became a symbol of the art and labor of the medical or culinary expert in spices, the cook and the pharmacist." Just as the mortar and pestle came to stand for the art of both the pharmacist and the cook, the word "recipe" in most languages came to mean, as Paul Freedman notes in *Out of the East*, "both instructions for cooks and prescriptions for druggists, a reminder of the conceptual similarity of the two professions." Nor were the "domestic arts" seen as lesser, or even as clearly divided from medicine and industry and "serious" endeavors. Home, like smell itself, had a status it has since lost.

So the Books of Secrets aspired to a weightiness and a comprehensiveness that might strike us as laughable in a contemporary "how-to." As Eamon observes, "We do not take very seriously the claim of the cookbook that professes to reveal 'all the secrets of the culinary art,' or the how-to book that promises to unveil the 'secrets of woodworking.' Such books may be useful, but few users will imagine you are going to learn more than how to make a tolerable meal or a sturdy piece of furniture. In the medieval and early modern eras, such claims carried much more weight than they do today. It seemed to many readers of the Book of Secrets that there was much more to be learned from a recipe than merely 'how to.'"

When I discovered the Books of Secrets, I was smitten both by their wealth of useful knowledge and by their charm. Their primordial scrambling of appetites and arts mirrored the synesthetic nature of the senses. Here home remedies mingled with folk wisdom, traditional knowledge with family lore. They combined the seemingly contradictory strands of the practical and the mystical in a way that reminded me of perfume, which—for all the formulas generated in all the labs in all the fragrance houses—can never be reduced to a science. Indeed, herbs, flowers, and spices played a great role in the arts they covered, and so the books focused on the same ingredients that are used to create natural perfumes.

I'd reveled in the same promiscuous muddling of material in my library of antique perfume books, in which fragrance recipes rubbed shoulders with alchemy, folk remedies, and precursor versions of aromatherapy. But when I tried to use the recipes to make perfume in a straightforward way—as if from a cookbook—I discovered something interesting.

I'd always thought, in the back of my mind, that if I ran out of ideas for new perfumes, I could use the recipes in these books to replicate the perfumes that used to be made—though I had not quite figured out how I would find, or afford, the copious amounts of musk and ambergris so many of the recipes called for. One day I decided to put that idea to the test and start making the perfumes in some of my old recipe books.

Many of the formulas were for what were known as soliflores, the attempt to replicate the aroma of a delicate flower that couldn't be scent-harvested, like lily of the valley or violet; they seemed to rely heavily on bitter almond, which smells like cherries, to convey the nuance of flowers. When I made a couple of the perfumes, however, they struck me as decidedly "old lady" and uninteresting. Moreover, as I combed through the books, giving the recipes a closer look, I could see that the recipes had frequently been copied from one book to another. I realized that the true genius of my antique perfume books lay not in their formulas but in the world they conveyed, a lost world of eccentric personalities consumed with the passion for travel to uncharted places, in search of undiscovered treasures and exotic substances—the same cast of characters who had inspired their pre-cursor Books of Secrets. The most important "secret" all of these books contained was an alternate way of looking at the world—an aura of romance, sensuality, adventure, and creativity.

As for the recipes in my antique perfume books, they taught me something, too—rather than how to replicate the perfumes of the past, I learned how to regard the process of making things and pass-ing on knowledge about that process. I felt that I was following an interesting, creative, admirable path that had been followed before. The recipes had the patina of having been forged in a crucible of trial and error by real-life practitioners who were handing down their hard-won knowledge to like-minded artisans. But recipes, even faithfully copied, cannot convey the intensely personal, idio-syncratic processes out of which they were distilled. "Recipes col-lapse lived experience into a series of mechanical acts that, once parsed, anyone can follow," Eamon observes. "While a 'secret' is someone's private property or the property of a group, a recipe doesn't belong to anyone. Once it is published, someone else appropriates it, uses it, varies it, and then passes it on. At each stop it gains something or loses something, is improved upon or degraded, and is changed to fit new needs and circumstances. Recipes are built upon the belief that somewhere at the beginning of the chain there is someone who does not use them."

Inherent in the Books of Secrets were attitudes and beliefs that grew out of the medieval imagination and resonated deeply with my work as an artisan perfumer. They reflected a belief in "maker's knowledge" (*verum factum*), which means that to know something means knowing how to make it. Such expertise cannot be acquired from someone else's experience but must be accrued by handling, experimenting with, and learning from the materials themselves. The search for how to do something was an essential part of the process, and it even had a name: *venatio*, "the hunt," which referred to the hunt after the secrets of nature. *Venatio* was exactly what I experienced when I was searching for the lost knowledge of natural perfumery.

The governing idea was that the "secrets" of a craft could not be taught by recipes or textbooks or in a classroom: they could only be learned through mindful experience with materials, a new journey every time. And the knowledge was cumulative: the learning wrought from mindful observation of experience led to the cunning and skill that produced more beautiful results. "The Greeks called this type of knowledge *metis*, by which they meant the kind of practical intelligence based upon an acquired skill, experience, subtle wit, and quick judgment: in short, cunning," writes Eamon. "Metis, or cunning intelligence, was entirely different from philosophical knowledge. It applied in transient, shifting, and ambiguous situations that did not lend themselves to precise measurement or rigorous logic."

But it is too simple to say that craft was therefore unscientific. "A distinction has often been drawn between the theoretical knowledge of scholars or scientists, which draws knowledge into a system, and practical craft knowledge, which is usually seen to be composed of a collection of recipes or rules that are followed more or less mindlessly," notes Pamela H. Smith in *The Body of the Artisan*. "Although there is a useful distinction to be made between theory and practice, investigation into the workshop practices of artisans belies such a view of craft knowledge." Art historians have discovered that artisans didn't merely collect and follow formulas but developed a profound, truly technical knowledge of the behavior and capacities of their materials.

The model for this experientially acquired expertise was alchemy, the ancient art that undertook to convert raw matter through a series of transformations into a perfect and purified form. Often referred to as the "divine" or "sacred" art, alchemy has complex roots that reach back into ancient China, India, and Egypt, but it came into its own in medieval Europe and flourished well into the seventeenth century. Paracelsus, the enormously influential sixteenth-century physician and alchemist, made an eloquent case for direct experience over received knowledge:

How can a carpenter have any other book than his axe and his wood? How can a bricklayer have any other book than stone and cement? How can a physician thus have any other book than the book that makes humans healthy and sick? Understanding must thus flow out of what one is, and the appearance of what one is must be tested.

Paracelsus ranked "doing" above "knowing" and thought that the knowledge acquired by doing was superior. "In such a way the physician recognizes medicines, the astronomer his astronomy, the cobbler his leather, the weaver his cloth, the carpenter his wood, the potter his clay, the miner his metal." As Smith explains, for Paracelsus, "knowledge of nature was gained, not through a process of reasoning, but by a union of the divine powers of the mind and the body with the divine spirit in matter. He called the process of this union 'experience.'" Science was acquired by uniting with nature in experience.

Artisans often engaged in alchemy and employed the language of alchemy to explain the transformation of materials and the redemptive dimension of their crafts. Both the processes and the metaphors of alchemy resonate with my own work. In the world of mass-produced perfumes, it is common practice to create a perfume without ever touching the essences. I taught myself to make natural perfumes by reading what the essences themselves have to teach me, and I continue to learn this way. I work in the most hands-on manner possible, using the essences at full strength, measuring with a dropper, mixing in small beakers, stopping to assess the changes in a blend after each addition. I see my work as giving voice to the materials I use. I often feel that the essences are my friends, each with its own distinct personality and spirit. I invest them with human qualities and discover their particular virtues by using them.

There is such a thing as artisanal understanding of the material world, learned from the manual labor of making something yourself, and the learning that comes though bodily knowledge of working with the materials. This knowledge comes not from writing and reading but by making and doing, and it is imprinted as much in the physical routines of the body as in the brain, the way a well-

mint oil for many centuries. Easy to grow and broadly useful in sweets and savories, beverages and foods, mint has also been recognized, everywhere it is grown, for its medicinal powers as well as its culinary uses. In Japan, mint was even thought to be effective as birth control. And in the Trobriand Islands, fresh mint is featured in a love charm: a would-be lover is to boil the herb in coconut oil as he chants a spell, then spill the (presumably cooled!) potion over his sleeping beloved's breast, inspiring her to have erotic dreams of him and upon awakening to find him irresistible.

Native Americans used mint widely in foods, drinks, cosmetics, and medicines. The Shoshone and the Paiute treated gas with mint tea; the Menominee treated pneumonia with a peppermint and catnip tea; the Lakota treated headaches with a tea from mint leaves or even the roots; the Blackfoot treated heart ailments by eating dried mint leaves; many Eastern Woodland tribes treated colds by inhaling the steam from boiling mint. Diluted oil of peppermint was sold by trappers and traders to many tribes; the distinctive clear, green, or blue glass vials with raised lettering in which it was sold have been found in many tribes' archaeological sites from the mid–eighteenth to mid–nineteenth centuries.

The mint family itself, Lamiaceae, is extensive, embracing a phenomenal number of species beyond peppermint and spearmint: basil, sage, rosemary, thyme, oregano, pennyroyal, lavender, perilla, chia, and even teak trees belong in the clan. Practically any of them might be what is meant by the all-embracing *yerba buena* ("good herb"), an abstraction common to the Spanish diaspora. In general, the term refers to whatever is the local species of mint, which varies from region to region and is seen as having healthful properties. The term covers a broad variety of aromatic true mints and mint relatives that may also have culinary value as teas or seasonings. In the Mediterranean the reference is to the traditional spearmint (*Mentha spicata*), while on the West Coast of the United States and Mexico, "yerba buena" describes kuntze or Oregon tea (*Clinopodium douglasii*). In much of Central America, it refers to bergamot mint (*Mentha citrata*); in Cuba it's usually *Mentha nemorosa* (also known as woolly mint); in Puerto Rico it's *Satureja viminea*, a close relative of savory; in Peru it's *Tagetes minuta*, a shrubby aromatic marigold, also known as huacatay or "black mint."

HAWKERS AND WALKERS

In early America the herb garden was an essential adjunct to the farmstead, typically featuring herbs that were easy to grow and had diverse uses: mint, marjoram, sage, parsley, and thyme. But in the New World, a homesteading ethos nurtured on an expanding frontier was also sustained by a new incarnation of an old character who is the opposite of a homebody: the peddler.

Settlers had to rely on their own skills and know-how. At the same time, peddlers made this self-reliance possible, by providing both materials that couldn't be grown or made and practical information and instruction on cooking, medicine, and more. Even in Colonial times, aromatics peddler was a recognized profession, as  distinct from, say, indigo peddler. "Usually a freelance," writes Richardson Wright in *Hawkers and Walkers in Early America*, "he managed to scrape together ten or twenty dollars, which was enough capital to set himself up in business, that is, fill his tin trunk with peppermint, bergamot, and wintergreen extracts and bitters." In that era every settler was a distiller, and the bitters were in great demand to mix with homemade spirits. Aromatics were also used in food and all kinds of home remedies.

> Here is another arcane recipe, for "Vienna Bitters," from the *Techno-Chemical Receipt Book*, 1896:
>
> Dissolve 40 drops each of oil of bitter oranges, oil of wormwood, and oil of Crete marjoram, 32 of oil of calamus, 20 each oil of peppermint, oil of marjoram, oil of anise seed, oil of thyme, and oil of cinnamon, 24 oil of coriander seed, and 12 of oil of cloves in 2 gallons of rectified spirit of 90 per cent. . . . Add 3 quarts of good red wine to the solution, sweeten it with 6½ pounds of sugar dissolved in 3½ quarts of water, color it red, and filter.

Peddling expanded with the frontier, and the peddler became a familiar figure there, his one or two small oblong tin trunks mounted on his back with a leather strap. There were the general peddlers who hawked an assortment of useful "Yankee notions"—buttons, sewing thread, spoons, small hardware items, children's books, and perfume. Bronson Alcott, Louisa May Alcott's father, left Yale to become a Yankee notions peddler before developing into a major figure of the transcendentalist movement.

Over time a peculiarly American subculture grew up around this nomadic subculture that included not only peddlers but also medicine shows, carny folk, fortune-tellers, dancing bears, minstrels, and all manner of "hawkers and walkers" who live on in our memory of what Greil Marcus has called the "Old, Weird America."

One pivotal figure in that world was "Dr." A. W. Chase. Born in 1817, he started out as a peddler of foodstuffs and medicines in Ohio and Michigan. For a while he traveled with the circus, collecting recipes—among them Backwoods Preserves, Good Samaritan Liniment, and Magnetic Ointment, which Chase insisted was "really magnetic" though it contained only lard, raisins, and tobacco—from the same people he peddled to: housewives, settlers, doctors, saloonkeepers. A recipe for Toad Ointment, a remedy for strain and injury that he got from "an Old Physician who thought more of it than of any other prescription in his possession," called for cooking live toads along with other ingredients. "Some persons might think it hard on toads," wrote Chase, "but you couldn't kill them quicker in any other way."

Eventually Chase settled in Ann Arbor, Michigan, where he printed a pamphlet of the recipes he had collected, giving it the title *Dr. Chase's Recipes; or, Information for Everybody.* This was a distinctly American Book of Secrets, and, like the one published by his predecessor Alessio Piemontese, it became a huge success, sold by peddlers much like himself to people who wanted a practical, all-purpose book to help them with a wide variety of daily problems. Over the next dozen years, Chase continued to add to it and to reprint it, until, by its thirty-eighth edition, it contained more than six hundred recipes. It was translated into German, Dutch, and Norwegian and sold all over the English-speaking world. Although he sold his rights to the book and to the printing house he had established, he ultimately lost his fortune and died a pauper in 1885. But his book lived on, selling about 4 million copies by 1915. According to William Eamon, "There were years when *Dr. Chase's Recipes* sold second only to the Bible."

Some of Chase's recipes were for things everyone needed—glue, ink, vinegar, ketchup—while others were specific to the needs of certain professionals, from bakers to gunsmiths. He organized it not by chapter but by "departments": "Saloon," "What and How to Eat," "How to Live Long," "What to Do Until the Doctor Comes," "Sheep, Swine and Poultry," and "Care of the Skin," to name but a few. His disquisition on vinegar captures the flavor of can-do exhortation that made his book such an enduring hit:

Merchants and Grocers who retail vinegar should always have it made under their own eye, if possible, from the fact that so many unprincipled men enter into its manufacture, as it affords such a large profit. And I would further remark that there is hardly any article of domestic use, upon which the mass of the people have as little correct information as upon the subject of making vinegar. I shall be brief in my remarks upon the different points of the subject, yet I shall give all the knowledge necessary, that families, or those wishing to manufacture, may be able to have the best article, and at moderate figures. Remember this fact—that vinegar must have air as well as warmth, and especially is this necessary if you desire to make it in a short space of time. And if at any time it seems to be "Dying," as is usually called, add molasses, sugar, alcohol, or cider—whichever article you are making from, or prefer—for vinegar is an industrious fellow; he will either work or die, and when he begins to die you may know he has worked up all the material in his shop, and wants more. Remember this in all vinegars, and they will never die if they have air.

Although experienced physicians regarded Chase as a charlatan, the medical remedies were the most popular aspect of his book. He recommends "soot coffee"—yes, made from "soot scraped from a chimney (that from stove pipes does not do)," steeped in water and mixed with sugar and cream—as a restorative for those suffering from ague, typhoid fever, jaundice, dyspepsia, and more. "Many

Instructions on how to deal with a burning woman, from *Harper's Universal Recipe Book*, 1869:

BURNING OF FEMALES BY THEIR CLOTHES HAVING CAUGHT FIRE.—A bystander, or the first person who is present, should instantly pass the hand under all the clothes, to the sufferer's shift, and raise the whole together, and closed over the head, by which the flame will indubitably be extinguished. . . . The sufferer will facilitate the business, and avoid serious injury, by covering her face and bosom with her hands and arms. . . . The females and children in every family should be told and shown flame always tends upwards, and that, consequently, while they remain in an upright posture, with their clothes on fire (it usually breaking out in the lower part of the dress), the flames, meeting additional fuel as they rise, become more powerful and vehement in proportion, whereby the bosom, face, and head being more exposed than other parts to this intense heat, or vortex of the flames, must necessarily be most injured: therefore, in such situation, when the sufferer is alone, and incapable, from age, infirmity, or other cause, of extinguishing the flames by throwing the clothes over her head . . . she may still avoid much torture, and save life, by throwing herself at full length on the floor, and rolling herself thereon.

persons will stick up their noses at these 'Old Grandmother prescriptions,' but I tell many 'upstart Physicians' that our grandmothers are carrying more information out of the world by their deaths than will ever be possessed by this class of 'sniffers,' and I really thank God, so do thousands of others, that He has enabled me, in this work, to reclaim such an amount of it for the benefit of the world."

A HOMEGROWN INDUSTRY

Many of the recipes in Chase's book and others like it were for toothpaste, mouthwash, gum—the very items that were beginning to be mass-produced outside the home, thanks to a burgeoning mint industry. Some of the early players in this native trade were the very peddlers who had catered to the needs of the homesteaders.

One pioneer in this industry, Archibald Burnett, helped expand commercial mint production from its original toehold in Ashfield, Massachusetts, into the "far West" area of central New York. On a trip peddling notions in 1810, Burnett ventured into the Finger Lakes region and found the land and people to his liking. He stayed to work on a farm and in 1814 married the farmer's daughter, whose name was Experience (yes, really!), and rented a nearby farm. His brother Nahum summoned him home to Ashfield to hear news so important that he "dared not trust even to the mail": Nahum had experimented with growing peppermint on his few stony acres, distilled the oil in a laundry tub, and turned a tidy profit. Archibald saw the opportunity to do even better growing mint in western New York and returned there with a bag of peppermint roots over his shoulder. Two of his brothers soon followed him, and as their farms flourished, many of the prominent peppermint growers from Ashfield migrated as well. Peppermint cultivation continued to expand in western New York and through Ohio and Michigan, and over the decades a competition for peppermint oil market share ensued between New York (known for higher quality) and Michigan (with larger production potential).

As Michigan gained in the mint rivalry against New York, an ambitious and innovative peppermint grower, Albert M. Todd, moved his center of production to Kalamazoo in 1891, starting experimental farms in the newly formed town of Mentha nearby. He hybridized plants collected on his European travels and used his education (he had a degree in chemistry) to improve distillation methods and increase his yield of high-quality refined oil. By the turn of the century, 90 percent of the world's supply of mint oil came from within a ninety-mile radius of Kalamazoo, as Todd's "Crystal White" mint oil became one of the most widely esteemed brands in the world. In 1907 a report in the *Farm Journal* featured Todd's operation as one of the country's agricultural wonders, comprising "more than 10,000 acres of land of which nearly one-half is devoted to the growing of peppermint, the remainder being utilized as pasture for many hundred short-horn cattle. In the winter these short-horns are fed 'mint hay,'—a by-product which remains after the oil has been distilled from the plants. So, you see, Mr. Todd makes 'two kinds of money' on the one product,—oil of peppermint money, and short-horn money; and everything is eaten up clean, except his profits!" Todd was known widely as the "Peppermint King," and several generations of his family continued to produce essential oils, but Mentha is now a ghost town—an empty train depot along an abandoned railway trail.

Long ago, soda fountains were opulent affairs, often in marble with mirrors and gold trimmings, and they rivaled saloons as meeting places. In 1911, according to *Fix the Pumps*, the United States had more than a hundred thousand soda fountains that served more than 8 billion drinks a year; now fewer than a hundred remain.

The fountain was presided over by the "soda jerk," whose job it was to squirt syrups into the sodas. Essential oils ranging from clove, orange, and peppermint to rose, musk, civet, and ambergris were used in the formulas, many of which seem playful and inventive. Created by pharmacists, they were often considered to be medicinal, and the recipes were kept secret. Coca-Cola, a product of soda fountains, was first formulated to treat dyspepsia and headache.

What really kicked the American market for mint essential oil into high gear was a product that at first didn't seem like much: chewing gum. The practice of chewing aromatic materials dated back to ancient times, and early American versions used a wide variety of flavoring bases, especially spruce oil. In the middle of the nineteenth century, the technique of manufacturing gum based on chicle, a natural rubber, was developed, and the gum business slowly began to grow. William Wrigley Jr., who had arrived in Chicago in 1891 with thirty-two dollars to his name and established a business distributing soap and baking powder, added chewing gum to his line. At first consumers didn't seem particularly interested, but then Wrigley's Juicy Fruit and Spearmint flavors took off, and by the end of the century, the industry was no longer growing slowly. Spearmint became the national favorite by 1910, to be supplanted only by Doublemint, flavored with peppermint oil, which Wrigley introduced a few years later. Nor was Wrigley the sole producer of gum, or mint-flavored gum. The wider acceptance of peppermint and spearmint essential oils in flavoring toothpaste soon followed, led by Colgate and Pepsodent among others, causing yet another spike in both the demand for and the quality of mint oils.

Weeds growing beside mint in the fields presented an ongoing challenge to that quality. When they were harvested along with it, they gave the essential oil a bad color or taste, or both. Growers in Washington's Yakima Valley stumbled upon a wonderfully down-to-earth solution to the problem: weeder geese, which ate the weeds but weren't attracted to the mint itself. One year an earthquake in the region shattered most of the goose eggs, and the resulting drop in the weeder geese population resulted in a corresponding increase in the quantity of weedy oil produced. What a poignant chain of effects, that a geologic event should affect the mintiness of toothpaste!

From such quaint beginnings was the flavor industry born. The history of deliberately flavoring food goes back thousands of years, of course. And the history of using essential oils for flavor is documented in cookbooks (and Books of Secrets) up to three centuries ago, in recipes for bitters, cakes, pudding, cordials, sauces, gum, cocktails, colas, ketchup, lozenges, candies, cookies, and more. But a distinct industry that sells flavor concentrates to food, beverage, and candy manufacturers is only around 160 years old. Their products were initially based on the natural extracts from herbs, spices, and resins that were already being used in the fragrance and pharmaceutical industries. As the methods of extraction became refined to allow isolation of individual aromatic components, they also yielded useful flavor materials. And as many of the fragrant mole-

cules began to be mimicked synthetically for the perfume industry, they likewise found their way into flavorings.

For more than a century, the flavor industry's ways remained almost entirely cloaked in mystery. Even the formulas
for "imitation" flavors composed from natural essences were
considered trade secrets worthy of protection. Eventually the
secrets of their formulations became widely known, largely
through the 1968 publication of the industry bible, *Food Flavorings: Composition, Manufacture, and Use*, by Joseph Merory.
It detailed some of the herbs and spices, essential oils, and synthetics
used to flavor common foods at that time. Here are examples of some
common foods and drinks and essential oils need to flavor them:

- ketchup—clove, cassia, nutmeg, pimento berry, mace,
  celery seed
- mayonnaise—black pepper, nutmeg, celery, lemon
- barbecue sauce—sweet marjoram, bay leaf, thyme, coriander, black pepper, pimento leaf, nutmeg, clove
- kosher dill pickle—clove, pimento berry, black pepper,
  dillweed
- sweet pickle—cinnamon, cassia, black pepper, coriander,
  caraway, pimento berry, clove
- pork sausage—black pepper, thyme, clove, ginger, bay
  leaf, nutmeg
- cola—lemon, lime, cassia/cinnamon, nutmeg, neroli,
  orange
- ginger ale—ginger, orange, lime, mace, coriander
- dry gin—juniper, lemon, bitter orange, angelica root,
  coriander, wormwood, cardamom, cinnamon, clove,
  mace, pimento, petitgrain, cognac

By the time *Food Flavorings* appeared, the Food and Drug Administration had recognized the maturity of the industry by giving it a kind of benediction, in the form of the Food Additives Amendment of 1958, creating the GRAS list of substances "Generally Recognized As Safe" and not required to be formally tested before being added to foods and drugs. The GRAS list, constantly updated as testing results become known, includes many essential oils as well as synthetic substances that can be used for flavoring. (All of the essential oils suggested for flavoring in this book are on the list.)

THE SWEET SMELL OF HOME

Today a familiar gesture of hospitality is the extending of a pack of gum (most likely mint-flavored) or a packet of breath mints. As a near-universal welcome mat that is often literally underfoot, mint has been closely entwined with the idea of hospitality—the art of making outsiders feel "at home" in a strange place by opening your home to them—for centuries. In ancient Egypt a Hebrew housewife gathered it from her garden to strew on her hard dirt floor, grinding

it with her heels to release its cool scent in an early version of "home fragrancing"—giving the home a welcoming aroma.

In a fable related by Ovid, Zeus and Hermes come down from Mount Olympus disguised as peasants and ask everywhere in town for food and a night's rest. After being turned away at a thousand wealthy households, they are finally welcomed at the humble house of an old married couple, Baucis and Philemon, whose preparation for the dinner included rubbing crushed mint leaves into the wooden table to fragrance it before serving their disguised guests a humble meal. Pleased with their treatment, Zeus and Hermes reveal their true identities to their hosts and take them to a hilltop just before a flood occurs. All the other houses in town are swept away except for the old couple's, which is turned into a lavish temple.

Guests in ancient Iran were given mint tea as a sign of welcome, a tradition that persists in parts of Asia and the Middle East to this day. A ritual experience, it is also laden with historical meaning. During the late nineteenth century, England eased trade restrictions on tea. Teatime became an egalitarian ritual, not just a privilege of the wealthy. And the tea trade, previously limited to the United States and Europe, spread to other regions. Moroccans began to combine Chinese green tea with mint and sugar, and their national drink was born. To this day Moroccans serve mint-flavored green tea before every meal, and always to guests, as a gesture of hospitality. I got to experience the ritual myself on the famous Marrakesh Express while traveling in Morocco in the 1970s. Amid the squawking chickens that roamed the swaying compartment, the tea seller would ceremonially—and steadily!—pour from a flimsy metal teapot crammed with spearmint and pieces of sugar that he hacked from a large hunk. Counterintuitive as it might seem in that sultry atmosphere, a steaming cup of tea tasting and smelling of the freshness of mint made me feel that all was right on that rickety train.

Just as fragrance can make us feel welcomed—can make a strange place feel like home—the wrong scent, or an unusually intense scent, can signal a place where something is not right. Daphne du Maurier does a pitch-perfect (or scent-perfect) job of using fragrance to evoke the eerie and not-quite-as-it-should-be atmosphere of Manderley, which stays with the nameless heroine of *Rebecca* long after the place has burned to the ground:

> A bowl of roses in a drawing-room had a depth of colour and scent they had not possessed in the open. . . . In the house they became mysterious and subtle. He had roses in the house at Manderley for eight months in the year. Did I like syringa? he asked me. There was a tree on the edge of the lawn he could smell from his bedroom window. His sister, who was a hard, rather practical person, used to complain that there were too many scents at Manderley, they made her drunk. . . . It was the only form of intoxication that appealed to him. His earliest recollection was of great branches of lilac, standing in white jars, and they filled the house with a wistful, poignant smell.
>
> The little pathway down the valley to the bay had clumps of azalea and rhododendron planted to the left of it, and if you wandered down it on a May evening after dinner it was just as though the shrubs had sweated in the air. You could stoop down and pick a fallen petal, crush it between your fingers, and you had there, in the hollow of your hand, the essence of a thousand scents, unbearable and sweet.

Every home has an idiosyncratic odor that we inhale when we first enter it. Wherever we have been, we recognize the unique smell of home as soon as we close the door and leave the outside world behind: the smell of rice steaming; your dog's breath on your face as he licks you; the musty smell of an old book in your library combined with the sharp, inky smell of a new one; your father's aftershave or pipe tobacco; your mother's shampoo. We value the aromas of home precisely for their contrast with the unfamiliar and exotic scents that greet our adventures. Scent is a kind of intimacy with place—our place in the world. Often we are unaware of the odor of our own home, because it is so familiar. But when we return there from afar, we can tell instantly if anything is "off" about it.

THE ONEIRIC HOME

For our house is our corner of the world. As has often been said, it is our first universe, a real cosmos in every sense of the word.

—GASTON BACHELARD, *The Poetics of Space*

Since the paradise of the world to come remains a mystery, the least and most we can do is to strive for an earthly one. Most of us find that place at home. Home is our mirror, our life story, and our peace. It's where we keep our books and treasures, commune with our friends, eat, sleep, and have sex. It's packed with memories, touchstones, and talismans—artifacts of our every age and all our travels.

My home is a typical Berkeley Arts and Crafts "brown-shingle" house built by the same unknown architect who designed the building that would become Chez Panisse restaurant behind me. The name refers to its exterior siding of unpainted wood shingles, which has darkened over the decades until it blends with the surrounding trees. Inside, there are Douglas fir floors, unvarnished redwood paneling, and large rooms with views of the Bay and the Golden Gate Bridge. My house has played a great role in inspiring my aesthetic and my perfumes, as has the surrounding environment. Here, I thought when I first came to Berkeley, was true Bohemia and a reverence for things that were handmade—not just homemade but embodying real craftsmanship and artistry, carrying with them the presence of the maker. The poet Charles Keeler had appreciated this ambience, too, when he moved to Berkeley almost a century before me in 1887. Recalling his introduction to the place, he wrote of the "delightful homes set in ample gardens with exotic flowers and near-by orchards." The effect must have been especially gorgeous without sidewalks or paved streets to interrupt it. Captivated by this unusual atmosphere, Keeler supported the building of homes by Arts and Crafts architect-designers Bernard Maybeck and Julia Morgan all over the city, and he became a leading national proponent of the movement. His adage that "home is the source of all art" certainly influenced my own story.

Leonard Woolf wrote that it is the house we live in that has "the deepest and most permanent effect upon oneself and one's way of living. The house determines the day-to-day, hour-to-hour, minute-to-minute quality, color, atmosphere, pace of one's life." The atmosphere of a house is like a vapor. When we feel in a cozy mood, we metaphorically draw it around ourselves like protection. In expansive times the house can feel as though it is opening up and blossoming like a flower.

The smell of home, or any smell we strongly associate with a particular person, place, or time, can bring on a flood of memories and powerful emotional responses almost instantaneously. These responses may be hardwired, but the particular smells that trigger such memories are a matter of conditioned response. As you first smell a new scent, you connect it to an event, a person, a thing, or even a moment. Your brain forges a link between the smell and a memory—associating lilies with a funeral, for example, so that when you encounter the smell again, the link is already there, ready to elicit a memory or a mood, sometimes without your even knowing why.

Thus our sense of smell is closely connected to our memories. Specific aromas are part of familiar rituals, seasonal celebrations, or heirlooms. The musty-dank smell of the basement, the yeasty smell of bread baking in the oven, the woodiness of a cedar closet mixed with the scent of mothballs and wool that signifies a change of seasons and therefore of clothing. Fragrances not only evoke powerful feelings of home; more than any other stimuli, they shape our intangible sense of the transformation of house into home. Home inhabits us, as much as the other way around; it lives in our imagination as well as on a city block or a country road, and partakes of the tonality of dreams and the poetry of loss. As Gaston Bachelard, matchless philosopher of the everyday, observes:

> If I were asked to name the chief benefit of the house, I should say: the house shelters daydreaming, the house protects the dreamer, the house allows one to dream in peace. Thought and experience are not the only things that sanction human values. The values that belong to daydreaming mark humanity in its depths. Daydreaming even has the privilege of autovalorization. It derives direct pleasure from its own being. Therefore, the places in which we have *experienced daydreaming* reconstitute themselves in the new daydream, and it is because our memories of former dwelling places are relived as daydreams that these dwelling places of the past remain in us for all time.

The poet Rainer Maria Rilke, having experienced this intimacy with the house of his youth, speaks of the fusion of being with the lost house:

> I never saw this strange dwelling again. Indeed, as I see it now, the way it appeared to my child's eye, it is not a building, but is quite dissolved and distributed inside me: here one room, there another, and here a bit of corridor which, however, does not connect the two rooms, but is conserved in me in fragmentary form. Thus the whole thing is scattered about inside me, the rooms, the stairs that descended with such ceremonious slowness, others, narrow cages that mounted in a spiral movement, in the darkness of which we advanced like the blood in our veins.

The lost house, the one that resides in our dreams—the oneiric home—reappears piece by piece from the foundation of our being. It is a shadowy recollection, like that of a scent. There is a fluid aspect to memories like this, an imprecise suffusion like the saturation of perfume on the skin and in the air. As Bachelard further observes, "It is as though something fluid had collected our memories and we ourselves were dissolved in this fluid of the past."

The significant home we have lived in, where we experienced intense feelings that became textured and emotionally suffused memories, finds its ghostly shape and substance in our imagination as a dreamy touchstone. We can return neither to that same building nor to the self we were there, and the memory of the house is like flypaper for all that was lost there. The oneiric home—resonant, unconscious, evocative, unknowable, poetic—cannot be pinned down with language; it needs to be experienced:

For the real houses of memory, the houses to which we return in dreams, the houses that are rich in unalterable oneirism, do not readily lend themselves to description. To describe them would be like showing them to visitors. We can perhaps tell everything about the present, but about the past! The first, the oneirically definitive house, must retain its shadows.

The walls, tables, chairs, and beds become imbued with the aroma of our day-to-day lives. That is why when we move into a new home, we often repaint or redecorate. We want to strip the house of the life that went on before we were there.

THE SWEET SMELL OF LUXURY

Material things have "biographies"—lives of their own. They "contain" us, beyond their being a commodity of routine usefulness. Our antique or heirloom objects have an appeal from the associations they evoke, either from dead family members or the imagined lives of previous owners. We can't separate the narrative of an object from the object itself; it is permeated with its story like an aroma. And it carries its story *in* its aroma. The past clings and enters us like a perfume, carrying worlds of feeling.

The lesson of home, the deep connection between material and meaning that it instills, is to be *better materialists*. "We are always assailing ourselves about being the most materialistic culture on earth," writes Richard Todd in *The Thing Itself: On the Search for Authenticity*. "Like almost everything else we say about ourselves, it is a half-truth." Our conspicuous consumption, he observes, masks a profound hatred of the stuff we buy. "We hate it. We want it to break and get dirty and wear out. We want to throw it away. And then, alas, we want to buy some more." Todd compares us as consumers to smokers who enjoy only lighting up a cigarette that they throw away after a few puffs. "I sometimes wish we were more materialistic, not less," he concludes, wishing that our culture needs "to love the physical, tangible world more than it does, to care more about what it creates, buys, places on the earth."

One of the hallmarks of modern consuming is the great differential between how we feel about an item before we make a purchase and how we feel after. Often, our strong desire for a thing evaporates once we've acquired it, and we discover that it means very little to us. Seeking to retrieve that feeling of desire leads us to accumulate large numbers of things that we lose interest in. Quickly.

In our modern technological age, we are especially prone to getting swept up in the "upgrade cycle," needing each component of our lives to match and interconnect. Once we have upgraded our kitchen appliances, can the countertops or floorboards be far behind? But this relentless consumerism, and its attendant striving for conformity, is not unique to our postindustrial age. In fact, it is known to today's consumer researchers as the "Diderot effect," after a famous story about the eighteenth-century French philosopher Denis Diderot (which he wrote about in an essay, "Regrets on Parting with My Old Dressing Gown") that Juliet B. Schor recounts in *The Overspent American*:

Diderot's regrets were prompted by a gift of the beautiful scarlet dressing gown. Delighted with his new acquisition, Diderot quickly discarded his old gown. But in a short time, pleasure turned sour as he began to sense that the surroundings in which the gown was worn did not properly reflect the garment's elegance. He grew dissatisfied with his study, with its threadbare tapestry, the desk, his chairs, and even the room's bookshelves. One by one, the familiar but well-worn furnishings of the study were replaced. In the end, Diderot found himself seated uncomfortably in the stylish formality of his new surroundings, regretting the work of this "imperious scarlet robe [that] forced everything else to conform with its own elegant tone."

One purchase seems to necessitate another, and another, and another. The deep pleasure of finding an item you love and bringing it into your life to be used with other things that you love seems to be a vanishing experience. If you buy a new jacket, you must get new pants; if you get new dishes, they will only go well with new silverware. Instead of welcoming a new member to the family of your belongings, you let it cast the older items in a shabby light. Conformity trumps individuality, erasing the long history of associations that makes our things precious to us.

As I asserted in the last chapter, some "luxuries" have been consistently costly—emeralds, for example—while others fluctuate with status, and some luxurious experiences are simple pleasures, like a bouquet of fragrant flowers, a piece of dark chocolate from Madagascar, a cup of fresh mint tea in a beautiful glass. Meanwhile, many so-called luxury items are shoddily made, with poor-quality ingredients. Coco Chanel is reputed to have said, "The best things in life are free. The second best are very expensive." The truth embedded in her joke is that price does not determine luxury: only quality does, and the associations that become embedded in a thing of quality.

Perhaps the antidote to overconsumption is simply to *buy less, better.* We could see our purchases as pieces of art rather than merely the next in a long line of replaceable consumer goods. The things we acquire become most precious when they become part of our memories and, in that sense, irreplaceable. By finding and using the best objects and materials—and by truly enjoying them, imbuing them with our own memories—we honor them. We bring them into our homes, and they become part of the fabric of our cherished memories. We fill our favorite teapot with a beautiful tea we've found and warm our hands on the special cups we've had for so long we can't remember where we got them; we bring a cup to someone we love and share our life with, and together we inhale the aroma of the tea before we take a first sip. In this way we make our things precious; we experience true luxury.

MINT VARIATIONS

Mint is the walking (or rather creeping) refutation to the idea that a substance must be expensive to be luxurious. Mint essential oils are very inexpensive, and you can find them almost anywhere that sells essential oils. It is among the most popular of the herbal aromas, and the most popular mint is peppermint, thanks to the high menthol content that makes it so invigorating. But spearmint, with a lower menthol content—only about an eightieth as much as peppermint—is milder and sweeter and more versatile for both scent and flavor.

To appreciate mint, perhaps more than with any other aroma, it is important to leave behind clichéd associations (toothpaste, chewing gum, and so on) in order to take in its live, fresh beauty with a "new nose." Try smelling several tiny samples, looking for the most pleasing, rounded versions. Good peppermint and spearmint oils have a sweetness to them, while those of lesser quality tend to be harsh and aggressive.

Also try comparing the essential oils with the aromas of the fresh leaves of different kinds of mint. Apple mint, a softer version of spearmint with fuzzy leaves, is sweet with apple overtones. Chocolate mint smells like a peppermint patty. Bergamot mint mixes a beautiful citrus with mint in one plant. There is a veritable cornucopia of lesser-known mints: water mint, corn mint, Japanese corn mint, American mint, pineapple mint, ginger mint, Bowles mint.

The mint family contains a number of other popular essences. Lavender essential oil is a top note that starts out herbaceous with a hint of eucalyptus, becoming more flowery as it evolves. As an absolute, lavender is a middle note that smells deeply of the blossoms; as a concrete, a base note, it has a flat, soapy smell. Basil essential oil, a middle note, has

A recipe for "Marseilles Vinegar," or "Four-thieves' Vinegar," from Arnold James Cooley's 1873 book *The Toilet and Cosmetic Arts in Ancient and Modern Times*:

It is said that this medicated vinegar was invented by four thieves of Marseilles, who successfully employed it, as a prophylactic during the visitation of pestilence. It is also said to have been a great favorite with Cardinal Wolsey, who always carried some of it with him as a preventative. This vinegar is a popular favorite on the Continent as a corrector of bad smells, and particularly of the air in sick rooms in fever, about the floors of which it is sprinkled. It was formerly in high repute as a prophylactic against the plague, fevers, and other contagious diseases. It has little pungency, but an agreeable and refreshing odor.

Take of
 dried rosemary 2 ounces
 oregano 2 ounces
 lavender flowers 1 ounce
 cloves (bruised) 1 drachm
 distilled vinegar 3 pints
and digest.

a sweet, spicy, faintly balsamic aroma, though there is a great range to the available basil oils, and some are sharp and without any sweetness at all. Tarragon essential oil is a top note with a sweet, green, anise-like aroma.

ODOR INTENSITY

Not all essences are created equal in terms of *odor intensity*, another important concept in scent literacy. Both peppermint and spearmint essential oils are top notes of high intensity. Knowledge of essences' odor intensities is required to understand how they interact in a blend. Higher-intensity essences can subsume the other notes, leaving a blend muddy, and can waste much money by obliterating the scent of an expensive but lower-intensity ingredient. A drop of high-intensity lemongrass, for example, can overpower the less intense absolute, whereas a jasmine absolute, with equally high odor intensity, could hold its own.

I like to think of relative odor intensity on a scale of 1 to 10—1 being the lightest odor intensity, a mere whisper of aroma, and 10 being equivalent to a person who dominates any gathering. A useful way to figure out how to balance various odor intensities, for both flavor and fragrance, is based on a method created by perfumer and perfume educator Jean Carles, which I have simplified here. To each of five small vials, add a milliliter of perfume alcohol or vodka. Add some of each essence to each vial, starting with a large amount of one and a tiny amount of the other, then decreasing one and increasing the other until, in the final vial, the proportions are reversed. Let's say we are doing an odor-intensity study of black pepper versus vanilla, for example. To the first vial, add 1 drop of vanilla and 9 drops of black pepper; to the next, 3 drops of vanilla and 7 drops of black pepper; then 5 drops of each, then 7 drops of vanilla and 3 drops of black pepper; and, in the last vial, 9 drops of vanilla and 1 drop of black pepper. Now put a drop of each mixture on a scent strip or just smell each vial. As the proportions shift, you will discover where the two aromas balance—their relative odor intensity—as well as the aesthetic sweet spot where you like the blend best, which is not necessarily the same thing. As a rule of thumb, in fact, I generally find that a 7/3 or 3/7 blend is more pleasing than a straight 5/5 blend.

An antique recipe for peppermint drops:

Stir 2 pounds of pulverized sugar into a stiff paste, with very little water, and then dissolve it in a well-tinned copper pan provided with a spout and a handle. Stir constantly, and let it become hot without boiling. Then add a few drops of oil of peppermint and pour a few drops from the spout of the pan upon a well-oiled metal plate. Should the mass in pouring prove too liquid, place it again on the fire and add more sugar. If it is desired to have the drops very strong, they are placed in a box, sprinkled with oil of peppermint, and the box closed as tightly as possible.

MINT IN FRAGRANCE

Mint presents a challenge in fragrance work because its high intensity needs to be invoked in an inventive way that creates a frisson rather than simply screaming *"Mint!"* I like to use a hint of mint to bring focus to a floral or a soft resin fragrance. It is also useful for punctuating green, earthy blends of oakmoss or woods, or to bring a fresh greenness to floral blends.

$\mathcal{M}$INT VETIVER PERFUME

I created this perfume to illustrate one way of dealing with spearmint's high odor intensity. Here the earthy, grassy aroma of vetiver grounds the mint and tempers its high register. Clary sage, with its amber notes, blends with practically anything, slides into the crevices between aromatics, and helps to smooth and marry them.

See pages 69–70 and 72–73 for basic blending procedures.

SOLID PERFUME

8 milliliters jojoba oil

Heaping ½ teaspoon grated beeswax

12 drops vetiver essential oil

9 drops clary sage essential oil

4 drops ylang-ylang extra essential oil

6 drops spearmint essential oil

OIL-BASED PERFUME

10 milliliters fractionated coconut oil

12 drops vetiver essential oil

9 drops clary sage essential oil

4 drops ylang-ylang extra essential oil

6 drops spearmint essential oil

BODY OIL

20 milliliters jojoba oil

4 drops spearmint essential oil

9 drops clary sage essential oil

12 drops vetiver essential oil

ALCOHOL-BASED PERFUME

8 milliliters perfume alcohol

BASE NOTES

6 drops vetiver essential oil

4 drops benzoin absolute

2 drops patchouli essential oil

MIDDLE NOTES

6 drops clary sage essential oil

6 drops ylang-ylang extra essential oil

TOP NOTES

4 drops spearmint essential oil

6 drops bergamot essential oil

2 drops black pepper essential oil

The benzoin and patchouli that make up the rest of the base in the liquid perfume is a particular combination that I love—the faint vanilla aromatic of benzoin paired with the aged cognac of patchouli lends an elegance to the sometimes dank and hippie smell of vetiver. In the middle notes, the clary sage helps amplify the solid green central chord. I like how the ylang-ylang softens the spearmint with its creaminess. Bergamot adds sparkle to the top notes, and the black pepper lends greater definition to both the spearmint and the entire opening of the perfume.

MINT IN FLAVOR

In flavor, spearmint essential oil, with its softer, rounder taste, is more useful than high-menthol peppermint. It's the menthol that's responsible for that zingy, cool-mouth sensation and makes peppermint so dominant, which tends to confine its use to candies and other sweets. Spearmint is equally at home in sweet and savory dishes and drinks. It's a perfect foil for both bland and spicy foods, and it adds a light freshness to fruit salads and lemonade. It pairs well with lime (think mojito), lemon (lemonade), chocolate, raspberry, strawberry, vanilla, cumin, cardamom, ginger, melon, tea, and lamb. Because of its intense and highly recognizable flavor, mint should be used with a light hand, to bring freshness rather than to overpower with mintiness.

Peppermint is best confined to ice cream, cocktails, cake, candy, and chocolate, but other members of the mint family have spearmint's versatility. Basil essential oil, like the herb itself, is incredibly useful in cooking, because the herb itself is seasonal and does not keep well. Like the fresh herb, it is perfect with tomatoes and in salads, and I love it paired with black pepper essential oil. Tarragon essential oil gives a gorgeous fresh herbal aroma to almost any vegetable.

An 1869 recipe for Crème de Menthe, from *Cooling Cups and Dainty Drinks:*

Dissolve in ½ pint of spirits of wine 1 drop of citron, 6 drops oil of mint; color green by a mixture of saffron and indigo; add a pint of syrup.

It is easy to make a half-ounce spearmint spray by combining 13 milliliters of vodka with 2 milliliters of spearmint. You can use this spray to flavor dark chocolate, lemonade, or vanilla ice cream or to scent black or green tea. Or you can make a mint tisane from fresh mint. Simply add a couple handfuls of leaves—the more varieties, the better—to a glass or ceramic teapot and fill with boiling water; steep, and serve. This tisane is also lovely mixed with lemon verbena leaves.

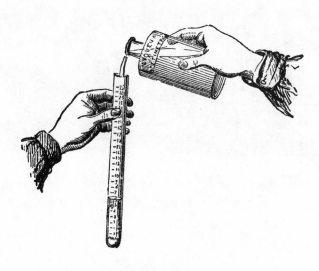

An engraving from André Thevet's 1575 La Cosmographie Universelle *depicts a group of men laboring to collect hardened frankincense resin from Boswellia trees.*

CHAPTER FOUR

REACHING FOR TRANSCENDENCE

Frankincense

Trees connect the three layers of the cosmos—the Underworld through their roots burrowing deep into the soil; the Earth's surface with their trunk and lower branches; the Heavens with their upper branches and top, reaching up to the light. Reptiles crawl among their roots; birds roost in their branches; and they relate the Under to the Upper World. They bring all the elements together: Water circulates in their sap; Earth becomes part of their body through their roots; Air feeds their leaves; and Fire is produced by rubbing their sticks together.

—JEAN CHEVALIER AND
ALAIN GHEERBRANT, *A Dictionary of Symbols*

THE ENCHANTING FOREST

With their roots in the ground, their limbs in the sky, and a life span that far exceeds our own, trees are a living symbol of the immortal. We feel this intuitively when we walk among them; standing under the canopy of ancient trees, inhaling their clean sweet air, we find it hard not to feel uplifted. Hermann Hesse captures this intuitive understanding when he recommends consorting with trees as wise advisers:

Trees are sanctuaries. Whoever knows how to speak to them, whoever knows how to listen to them, can learn the truth. They do not preach learning and precepts, they preach, undeterred by particulars, the ancient law of life. . . . When we are stricken and cannot bear our lives any longer, then a tree has something to say to us: Be still! Be still! Look at me! Life is not easy, life is not difficult. Those are childish thoughts. Let God speak within you, and your thoughts will grow silent.

The sacredness of trees is universal. The Bible has its Tree of Knowledge, the Kabbalah its Tree of Life; Buddha received enlightenment sitting under a tree. Sometimes the tree has stood for the process of the psyche becoming conscious and individuating. In many myths people turn into trees, and vice versa. In many countries there is a tradition of planting a tree to commemorate the birth of a child. Trees can represent a better version of being, a symbol of human transcendence.

Trees also represent the deepest spiritual connection of which humans are capable. Remember Baucis and Philemon, who showed such hospitality to the peasants who were really Zeus and Hermes in disguise? In addition to turning their home into a golden temple, the gods wanted to know what more the old couple wished as reward. They asked to be priests of the temple while they lived out their years, then to die together. So they tended the sanctuary until one day, as they stood before the temple, recalling how it had been created, each saw that the other had begun to sprout leaves and branches. They called farewell as they grew into entwined trees, a twin oak and a linden tree that shared a single trunk.

It is in the collective that we experience the power of trees most profoundly. Standing together in a forest, they form an intensely altered space, creating interiors in the outdoors. We sense an edifice, roof and walls, hidden hallways and distant corners, a series of rooms and galleries. And not just any rooms or galleries: forests are like temples. Or, more accurately, temples—places of worship—are like the forest. As John Fowles muses in *The Tree*:

> I am certain all sacred buildings, from the greatest cathedral to the smallest chapel, and in all religions, derived from the natural aura of certain woodland or forest settings. In them we stand among older, larger and infinitely other beings, remoter from us than the most bizarre other non-human forms of life: blind, immobile, speechless . . . waiting . . . altogether very like the only form a universal god could conceivably take.

In the seemingly unlimited world of the forest, we experience vastness, much as we do when we look out at the sea from shore. Once we venture into the forest, we cannot see beyond the veil of trunks and leaves, but we sense the infinite journeys that lie down paths in any direction. The experience of the forests can't be fully captured in words or images, because it is made up of the richness of aromas, the mood of light, the texture of air, the drama of upward motion and our place inside that verticality. As we stand in the woods, we feel viscerally the diversity and immensity of nature.

In folklore and fairy tales, the forest is a place of enchantment and change, a setting for magic, both good and ill. To enter there is to cross a threshold to significant transformation. In folklore the danger that lies in the forest is an opportunity for the hero. One of the oldest of all recorded tales, the Sumerian epic *Gilgamesh*, recounts how the hero and his companion, Enkidu, travel to the Cedar Forest to fight the monsters there and to be the first to cut down its trees.

Hansel and Gretel, and Snow White, venture into the woods and emerge transformed. As Fowles elaborates, "It is not for nothing that the ancestors of the modern novel that began to appear in the early Middle Ages so frequently had the forest for a setting and the quest for central theme. . . . The metaphorical forest is constant suspense, stage awaiting actors; heroes, maidens, dragons, mysterious castles at every step."

A GODLY STEAM

Given the mysterious aura of trees, it seems fitting that the aromatic materials that most profoundly galvanize the soul come from them. The first perfume was incense, its earliest uses bound up in religious rites. Incense is made primarily from aromatic woods and their various gums (polysaccharides) and resins (hydrocarbons), which are the thick, ambrosial syrup that runs in the veins of certain trees and shrubs: frankincense, myrrh, labdanum, dragon's blood. Although these aromatic resins appear to be useful in repelling the plants' enemies, they often have the opposite effect on humans. With their complex aromatic layering, these substances are a vehicle for spiritual exaltation. Mixed with spices and burned, they release a sweet, penetrating aroma that transports us. The burning of aromatics opens a door between the mundane and the supernatural, an elevation of consciousness that has been incorporated into the rituals and belief systems of many religions. Indeed, incense has permeated spiritual practice as thoroughly as spiritual practice has permeated human life. An experience of scent heightened by the earthiest of materials gives a rocket boost to the flight of transcendence.

The white smoke of frankincense was the ultimate balm for angry gods of whatever persuasion: one temple in Babylon burned two and a half tons of it a year. But evidence suggests that aromatics came into ritual use initially for practical reasons: the burning of fragrant wood masked the smells of cremation and human and animal sacrifice, older forms of offering. Ancient Hindus sacrificed cows, goats, sheep, water buffalo, horses, and men. Along with the flesh, they burned tufts of wool and dung. They believed the spirits of the dead to be sustained by the aroma of burnt offerings, and they used the resulting ashes to anoint themselves. Incense acquired its religious significance by association, in other words: the burning of incense replaced the blood sacrifice it once accompanied.

Frankincense was not only burned by the Egyptians but used in embalming. Copious amounts of it and other fragrant materials were packed along with as well as inside the dead. Priests closely guarded the recipes for these highly potent unguents. When King Tut's tomb was opened after three thousand years, jars of such unguents were found within—still fragrant—along with pellets of frankincense.

The popularity of incense was driven by the spread of religion. In the first century A.D., Buddhism moved from India into the rest of Asia, bringing with it a much more intense and varied use of incense. Aromatics were already in use in China, and there, as elsewhere, their ritual applications were not cleanly distinguished from their medical, cosmetic, and culinary employment. "These were burned to produce a pleasant fragrance to be used on the person while bathing, on clothing, in places of worship, in the home, and in government offices," especially for business involving the emperor, notes Silvio A. Bedini in *The*

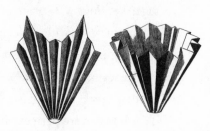

Trail of Time. Incense was considered essential not only for attracting the favor of gods and lovers but for establishing and sustaining a pleasant frame of mind: for well-being.

The expansion of Asian trade brought an influx of new aromatics. Ships arrived laden with sandalwood and aloeswood (also called agarwood or *oud*), camphor and patchouli, benzoin and storax, frankincense and myrrh. "The acknowledged superiority of the Indochinese aromatics, beside which those of China were 'beggar's incense,' and the apparently inexhaustible sources of perfumes and incenses from 'groves whose rich trees wept odorous gums and balms' in the vaguely defined 'South Seas,' gave rise to the idea of a kind of incense tree, which bore all the important aromatics together," notes Edward Schafer in *The Golden Peaches of Samarkand.* "Its roots were sandal, its branches were aloeswood, its flowers were clove, its leaves were patchouli, and its gum was frankincense." The tree may have been mythic, but the fetish for incense was not. China—Canton in particular—became a hub of the trade. Much of it was consumed domestically, by the aristocracy, who even scented architectural structures. According to legend, a revolutionary built himself a palace entirely of incense woods, to which he set fire in protest, perfuming the countryside for miles around.

In China and elsewhere in Asia, incense had long been used for rituals relating to the afterlife: it was burned in the presence of the newly deceased, to protect the body and to ease the soul's passage, keeping evil spirits at bay. There was a special incense for communicating with the dead, known in Chinese as *fan hun hsiang* ("calling-back-the-soul aromatic") and in Japanese as *hangon-ko* ("spirit-evoking incense"). Buddhism took the use of incense to another level, binding it into every aspect of observance. "Buddhist books were permeated with aromatic images, and indeed the

Sanskrit word *gandha*, 'aromatic,' often means simply 'pertaining to the Buddha,'" observes Edward Schafer. "A temple was called *gandhakutī*, 'house of incense'; the pyre on which the Buddha was cremated became a 'fragrant tower'; 'Fragrant King' and 'Fragrant Elephant' were epithets of Bodhisattvas; and on Gandhamādana, 'Incense Mountain,' dwelt the *gandharvas*, gods of fragrance and music." So key was the role of incense in religious practice that dishonest trade in it was thought to doom the culprit to a specific kind of hell: "According to Japanese legend," Bedini writes, "any form of burning incense will summon the *jikikōki* or 'incense-eating goblins,' a class of hungry demons (*pretas*) recognized in Buddhism. These are the souls of men who in their lifetime had been guilty of selling incense of inferior quality for profit, and who are now compelled to seek their only food in incense smoke."

In the West, too, the ritual use of incense evolved out of its practical uses, and it is difficult, retrospectively, to say where one left off and the other began. The streets of ancient Rome were awash in the stench of rotting meat, excrement, and rubbish. They were also punctuated with statues of the emperors, each of which was accompanied by its own incense burner billowing fragrant frankincense-laden smoke, to counteract the noxious odors. Incense was offered on domestic shrines and in temples, at weddings and sporting competitions and cremations. In 78 B.C., the funeral of the dictator Sulla featured a life-size statue of the departed, fashioned from frankincense and other aromatic materials, which was burned, emitting fragrant smoke throughout the ceremony.

Sacrifices were ritualized occasions steeped in aromatics from beginning to end. If there was a procession, incense was burned along the way. The participants and the statuary were festooned with sweet-smelling wreaths. The wine imbibed was infused with

 flowers and spices. If an animal was being sacrificed, there was a strong base note of roasting meat. "A godly steam, and fit for godly nostrils, rises heavenwards, and drifts to each quarter of the sky," as the second-century rhetorician and satirist Lucian described it. The scent of sacrifice was complex and variable. "It might be as simple as frankincense alone," writes Susan Ashbrook Harvey in *Scenting Salvation*, or "it might carry the grand fragrances of extravagant ceremony." But the burning and consuming of incense became more than an important accompaniment to sacrifice; they became a symbol, and then the reality, of sacrifice itself.

A "Universal Wound Balsam" from *Dixon Encyclopedia*, 1891:

Gum benzoin, in powder, 6 ounces; balsam of tolu, in powder, 3 ounces; gum storax, 2 ounces; frankincense, in powder, 2 ounces; gum myrrh, in powder, 2 ounces; aloes, in powder, 3 ounces; alcohol, 1 gallon. Mix them all together and put them in a digester, and give them a gentle heat for three or four days; then strain. 30 or 40 drops on a lump of sugar may be taken at any time, for flatulency or pain at the stomach; and in old age, where nature requires stimulation. This valuable remedy should be in every family ready for use; it cannot be surpassed as an application for cuts and recent wounds, and is especially good for man or animals.

The entwining of aromatic substances with the impulse toward transcendence is perhaps most vividly illustrated by the legend of the phoenix. What does the mythological bird feed upon? "It does not live on fruit or flowers, but on frankincense and odoriferous gums," writes Ovid. And "when it has lived five hundred years, it builds itself a nest in the branches of an oak, or on the top of a palm tree. In this it collects cinnamon, and spikenard, and myrrh," and it is from the ashes of these sweet substances that the great bird rises again.

The burning of frankincense and other aromatic resins—and its spiritual significance—continued uninterrupted into the Judeo-Christian tradition. In a milieu in which spices and perfumes were in common use and carried a variety of cultural meanings, the rabbis of the late Roman and early Byzantine periods bent them to their own theological purposes—for example, developing their own blessings for the burning of incense.

In the Bible, olfaction is one of the primary modes of interaction and communication with God. Each properly performed sacrifice was conceived as "a soothing odor before God." In Genesis, when Abraham circumcises himself and his sons and his sons' sons, he makes a hill of their foreskins. The aroma of the decomposing foreskins, which the Bible likens to that of burning frankincense, rises to heaven, where it is pleasing to God, focusing his attention on earth and transforming his mood:

And when the sons of this man come [to do] transgressions and evil deed, I will remember for them this smell and I will be filled with mercy for them. And I will convert for them the Attribute of Justice to the Attribute of Mercy.

Moreover, sacrifice is a two-way street. We burn incense—in fact, the name is essentially synonymous with burning: "incense" comes from the Latin *incendere*, meaning "to burn." But we also breathe in—*spirare*—what we burn, and are thus inspired. And we do so collectively, as a congregation. The sweet smells that men produce to please God are holy, sanctified; they are the divine in us. In Exodus, God orders Moses to have a perfumer blend a special anointing oil of myrrh, cinnamon, calamus, and cassia in an olive oil base, to be offered on his gold-covered altar. He is to have incense compounded, too, from the resins storax, frankincense, and galbanum, along with onycha, a mollusk shell. "Most holy shall this incense be unto you," God proclaims, prohibiting the mixtures from being used for any purpose except worship.

What else would aromatic smoke coiling toward the heavens be, after all, if not a pathway to God? Scent, in a sense, *is* spirit: potent, invisible, omnipresent, elusive, capable of transforming experience or meaning. It is easy to see why it was believed that sensory experiences carried cosmological significance, ordering the cosmos. The good smells good: beautiful smells partake of the divine. The Bible's most secular verses, the Song of Solomon, are famously replete with sensual images of wafting perfume and seductive spices. "While the king was at his repose, my spikenard sent forth the odor thereof. A bundle of myrrh is my beloved to me, he shall abide between my breasts," reads one passage, and another: "My beloved put his hand through the key hole, and my bowels were moved at his touch. I arose up to open to my beloved: my hands dropped with myrrh, and my fingers were full of the choicest myrrh." It is not just that no distinction is made between the sexual and religious ecstasies that intense olfactory experience inspires. It's that the experience of the godly is marked by the most heightened sensory experience of which we can conceive. As Paul Freedman elucidates the above passages:

Christian interpretations considered the constant evocation of perfumes to be a sign of divine presence. The third-century Alexandrian theologian and commentator Origen said that the bridegroom in the song [Song of Solomon] was Jesus and that the bride and the maidens run after him because of his holiness, symbolized as fragrance. The divine scent demonstrates something that can be tangibly felt, but not by the common senses of sight, hearing, or touch. The sacred aspect of olfactory sensation was reinforced by its unseen intensity.

The ultimate Christian parable about the sanctity of scent is the story of the elderly bishop Saint Polycarp, who was executed in Smyrna in A.D. 155 because he refused to sacrifice to the Roman gods. As an eyewitness account has it:

The men in charge of the fire started to light it. A great flame blazed up and those of us to whom it was given to see beheld a miracle. And we have been preserved to recount the story to others. For the flames, bellying out like a ship's sail in the wind, formed into the shape of a vault and thus surrounded the martyr's body as with a wall. And he was within it not as burning flesh but rather as bread being baked, or like gold and silver being purified in a smelting furnace. And from it we perceived such a delightful fragrance as though it were smoking incense or some other costly perfume. At last when these vicious men realized that his body could not be consumed by the fire they ordered a *confector* to go up and plunge a dagger into the body. When he did this there came out such a quantity of blood that the flames were extinguished.

The story is redolent with smells and their symbolism: the aroma of baking bread, with its ties to daily sustenance and its relationship to the body of Christ; the aroma of frankincense, the quintessential "soothing smell" of sacrifice pleasing to God.

Indeed, beautiful smells are not only offered up to God but emanate from the divine. It was a basic precept that heaven was filled with beautiful smells and hell radiated a putrid stench. Paradise smelled sweet, and good smells were understood as the vapors exhaled by a world beyond. Extrapolating from this, every foul or beautiful odor was a meaningful foretaste of the world to come and underscored the sense of earthly life as a mere prelude to the afterlife.

And the sense of smell itself was seen as uniquely incorruptible among the senses, uniquely capable of exaltation. One of the most famous passages in the Old Testament describes how the Messiah will use his incomparable powers of perception to discern the good:

> *And by his smelling in awe of the Lord,*
> *And not by [what] his eyes see, will he judge,*
> *And not by [what] his ears hear, will he decide.*
>
> —Isaiah 11:2–3

This passage points to the supreme wisdom of "smelling things out." Even God's chosen one may be misled if he trusts the evidence of his eyes and ears, but he will choose rightly if he follows his nose. He will be able to see past and hear through lies and deception and to sniff out the truth. "In the overall context of the passage, the idea of this chosen person 'smelling' . . . the truth means that he will sense it, discern it, or otherwise feel it in some manner," Deborah Green points out in *The Aroma of Righteousness*. "It demonstrates that the concept of 'smelling' as a form of intuition or acquiring knowledge is of ancient pedigree. It also confounds the approach of Western philosophy and psychology, which devalue this sense in comparison with the 'higher' senses of sight and hearing." In holy terms, just as that which is good smells good, those who are good smell well.

A GIFT FROM GOD

As many as twenty-five species of the *Boswellia* tree produce the gum resin frankincense, highly prized by almost all the peoples of the Middle East and the Mediterranean for its pungent, pleasurable smoke. As with cinnamon, its sources were often protected by the usual dramatic obfuscation: Herodotus retells a tall tale that local tribesmen told to gullible foreigners, about flying snakes of various colors that kept guard over the trees and had to be smoked out with burning storax, an aromatic gum.

Boswellia sacra, historically the most prized of the frankincense trees, originated in the remote, inhospitable Dhofar region of Oman, famous for its high-quality "hojari" frankincense. There the trees were regarded as a gift from God and were allowed to grow wild rather than being cultivated. But *Boswellia* trees were widespread across the Horn of Africa (Somaliland) and farther east in Arabia, and no one people had a monopoly on the frankin-cense trade: the Minaeans and other peoples of the Arabian Peninsula had a share in it, including the Qedarites, the Gerrhaeans, and the Nabateans. At first, primitive donkey caravans transported the merchandise, but they could carry only small loads for short distances. By 900 B.C. the single-humped

Arabian camel (*Camelus dromedarius*), much better adapted to long-distance transportation, had been domesticated, and it proved a boon to the frankincense trade.

Alexander the Great was apparently a frankincense junkie. Plutarch recounts that when the future world-beater was a boy, his tutor chided him for the excess of his offerings: "When you conquer the lands where these sweet things grow, then be extravagant with this perfume." Indeed, Alexander would eventually conquer Gaza, the center of the incense trade, cornering the market on this temple staple. As legend had it, he sent five hundred talents (about nineteen tons) of frankincense and a hundred talents of myrrh to his old teacher, with a note: "We send you plenty of frankincense and myrrh so that you no longer need be stingy with the Gods!"

To harvest the frankincense, the bark was shaved, creating wounds in the trunk from which resin oozed. When it hardened into crystals, they were scraped off the tree. Frankincense is sometimes called olibanum, from the Hebrew *lebonah*, meaning "milk." This is a reference to the appearance of the freshly exuded resin. In the same vein, Pliny noted that the most prized frankincense of all was that which was breast-shaped, a form produced by one drop of resin hardening upon another and another.

The process of harvesting remains much the same today, and the most common varieties are from the *carterii*, *frereana*, and *thurifera* species of *Boswellia* from Somalia, the *papyrifera* and *rivae* from Ethiopia, and the *serrata* from India. The freedom to buy freshly distilled frankincense directly from indigenous distillers is a relatively recent development, and I have loved discovering them all. My favorite frankincense essential oil comes from the underrated *Boswellia neglecta* from Kenya. Even the supplier, an organic grower, often asks me if I would prefer a more popular variety. But I love the abundant sparkling citrus and coniferous top notes with which it opens, followed by soft, mellowing balsamic notes, without any trace of sharpness.

In the regions that grow frankincense, it is still deeply woven into medicinal practice and religious and social rituals. It is burned for forty days following the birth of a child and is thought to prevent infection. It is used to flavor coffee and to treat dental problems, arthritis, and coughs. Women chew the resin to sweeten their breath and use the oil to groom their hair. People fumigate their homes with it, and women use the resulting soot as kohl, to line their eyes. Pots are mended with the heated resin. Frankincense is believed to promote peace and harmony and is therefore burned at weddings and religious celebrations. A frankincense-scented candle is believed to keep away evil spirits. Like that of mint, the aroma is also regarded as a sign of welcome.

Recent research lends scientific support to these millennia of frankincense use. Researchers at Johns Hopkins University and the Hebrew University of Jerusalem have confirmed the psychoactive properties of frankincense. They have determined that frankincense smoke relieves depression and anxiety in mice and have isolated the chemical compound—incensole acetate—that is responsible for these effects. Frankincense may also hold a clue to a cure for cancer. According to recent studies, it "separates the 'brain' of the cancerous cell—the nucleus—from the 'body'—the cytoplasm, and closes down the nucleus to stop it reproducing corrupted DNA codes." Research also substantiates the longtime use of frankincense in treating arthritis. Scientists at Cardiff University in Wales believe they have been able to demonstrate that treatment with an extract of frankincense inhibits the production of key inflammatory molecules, helping to prevent the breakdown of cartilage that causes the condition.

It's impossible to talk about frankincense without mentioning its forever companion, myrrh, another aromatic resin. In the fifteenth century B.C., the Egyptian queen Hatshepsut dispatched a royal expedition consisting of five ships to the Land of Punt (thought to be part of modern-day Somalia) to bring back myrrh and myrrh seedlings to plant in her temple. Punt was a voyage of about two thousand miles across the treacherous waters of the Red Sea. The expedition was commemorated in inscriptions and reliefs on the walls of her temple, depicting the precious cargo of apes, monkeys, dogs, and panther skins, along with the myrrh.

Egyptians used frankincense, myrrh, and other prized resins almost daily. "The Egyptian word for myrrh, *bal*, signified a sweeping out of impurities, indicating that it was considered to have medicinal and, ultimately, spiritual properties. Ancient Egyptians scented their homes and were commanded to perfume themselves every Friday. Idols were regularly anointed with perfumes, and incense was an important element in religious ceremonies."

Another illustrious incense ingredient, aloeswood, is the most expensive perfumery material in the world, with some specimens in Japanese museum collections costing six figures. This precious aromatic wood comes from pathologically diseased parts of *Aquilaria* trees native to Southeast Asia; when a tree is infected with the naturally occurring fungus *Phaeoacremonium parasitica*, its immune response produces an aromatic resin to suppress the fungus, making that section heavier and darker than the surrounding wood. The resin can be extracted through steam distillation of the wood, with 150 pounds of wood yielding less than an ounce of essential oil. In the wild forest, the infection occurs in fewer than 10 percent of the trees, while some commercial forest operations deliberately inoculate each tree.

After the afflicted trees are chopped down and left to rot, the rotted surface is removed, revealing the dark, odoriferous heartwood. Sometimes the pathologically fragrant patches occur in the shapes of men or animals, which increases their market value greatly.

Aloeswood (*oud*) has been used in incense and perfume since ancient times, mentioned as far back as the Sanskrit Vedas from India. It was also tinctured in wine and used in Chinese medicine for internal pains and external lesions, and was deemed powerful enough to purify the soul and drive out evil spirits. Courtesans' garments were sometimes soaked in *oud*-infused water, presumably to enhance the women's sensual appeal. One nobleman found an even more decadent and extravagant use for the precious wood: He had it made into a powder and applied to the walls of his mansion, so that visitors would be overwhelmed by the scent when the door was opened.

As with other rare imported goods, the source of this valuable material was historically safeguarded as a treasure in itself, by both the Malays and the merchants who profited so greatly from its trade. It helped that Malay, rich as a spoken language, had no indigenous written form in which the secrets of *oud* could be divulged. In recent times *oud* has become the new "it" perfume ingredient, perhaps as a backlash against the multitude of "fresh and clean" perfumes churned out by the mass-market fragrance industry over the last many years. It is very difficult to describe its unique, complex appeal, at once sweet and sour, spicy and animalistically rich, earthy and woody.

Apart from the heavenly fragrance of these substances, among the delights of getting to know and use them is the experience of continuity with our long history that they bring us. So little has changed over thousands of years even in the way they are harvested. Frankincense still grows wild in Oman. Its bark is still scraped to create the precious resinous "tears." Local people still use it as they have always done. Smelling and using it ourselves, we are not so far removed from those three ancient kings who crossed the desert carrying precious resins in their intricately woven saddlebags.

LISTENING TO INCENSE

By burning incense we know the o'clock of the night.
—YU CHIEN-WU, *sixth-century Chinese poet*

In Japanese, the word for "to sniff" is the same as the word for "to listen." This linguistic coincidence led to the development centuries ago of a ritual ceremony, *kōdō* (way of incense), still so popular that it served as the inspiration for a 1987 work called *Kō o kiku* (*Listen to the Incense*) that had its world premiere in a Boston Symphony Orchestra performance conducted by Seiji Ozawa. In the traditional *kōdō* ceremony, noblemen, warriors, sages, and priests gathered to "listen" to incense with "ears of the spirit." Incense burners containing chips of each of three different kinds of aromatic woods were passed around, with the kind of wood labeled on them, and the participants commented on and memorized the characteristics of each. Then one kind of wood chip was chosen and that incense burner was passed around again. But this time its name was hidden and the participants had to guess which fragrance it was. In other words, they first "sniffed" and then "listened to" the incense, translating their responses into poetry for the final judging. The point of the game was to teach the art of detecting the nuances of different kinds of incense and, more broadly, "to calm the mind, and to constitute an effort to find meaning in a materialistic world while engaged in communication with friends."

Just as aromatics that were burned alongside the sacrifice became, in effect, the sacrifice itself—material transubstantiating into symbolic ritual—so did the ritual of incense take on a life of its own, adapted to all kinds of sensual, practical, whimsical, and fantastical purposes. In the Far East, for example, incense was used to mark time, as the combustible base materials were refined enough to burn at a consistent rate. Compressing the incense into a long stick form (with or without a bamboo core) made it especially suitable for timekeeping: the burning time for a given kind of stick was noted, and then time segments could be marked along the length of individual sticks so that when they burned, the passage of time could be easily noted by the progress of their burning. Around the eighth century, the Chinese developed a highly sophisticated form of incense clock. Powdered incense was spread on a flat surface and incised with characters standing for the divisions of time. This was called an "incense seal" or "aromatic seal," because the figures etched in the incense were similar to those carved on a gentleman's personal seal. When the incense was lit, time could be tracked as the fire burned its way along the narrow trail of incense that led from sign to sign. Not to be outdone, the Japanese further refined the incense seal, creating incenses with distinct aromas to mark off twelve-hour intervals.

Even when mechanical clocks and watches were introduced to China from the West, they were not widely adopted. The incense clocks had a revered home on the scholar's desk, and as their function was not only utilitarian but inspirational, they were not easily displaced. In fact, the precise time of day was apparently not a matter of urgent concern in nineteenth-century China, as observed by American residents there: "Clocks in public buildings, particularly railroad stations, appeared never to register the time with any degree of accuracy," notes Bedini in *The Trail of Time*. "When one of these public clocks malfunctioned, no attempt was made to repair it; instead a new clock was added to the wall and the inoperative ones allowed to remain. Travelers noted seeing walls of some public offices and railroad stations covered with a number of clocks, most of them no longer functioning, the others inaccurate."

Bedini also tells the story, even more poetic and imaginative, of a nineteenth-century French missionary, Père Evariste Régis Huc, who came upon this practice of time-telling in rural China:

One day as he was passing through a rural region with several companions, he stopped to ask a farm boy whether it was yet noon. The sky was overcast and the sun was hidden by clouds. The boy asked them to wait, left his work and returned moments later holding a cat in his arms. No, he reported, it was not yet noon; one could tell from the cat's eyes. Unwilling to display their ignorance, Huc and his companions accepted the information, without further questioning the boy. They thanked him and continued on their travel.

Later, when Père Huc and his companions had an opportunity to inquire of others, it was explained to them that observing a cat's eyes to tell the time was a common practice among the rural Chinese. They were aware that the pupil of the cat's eye kept diminishing more and more as daylight increased. By noon the pupil was reduced to a very thin line, and then the pupil gradually dilated and became larger and larger as the afternoon advanced.

Chinese sources stated, "The pupil of the cat's eye marks time; at midnight, noon, sunrise and sunset, it is like a thread; at 4 o'clock and 10 o'clock, morning and evening, it is round like a full moon; while at 2 o'clock and 8 o'clock, morning and evening, it is elliptical like the kernel of a . . . date."

Using a cat's eye to tell time seems like a dreamlike image out of Bachelard's oneiric house. When you can dream up and use something as poetic and romantic as an incense clock, why even consider using an ordinary clock?

And perhaps that is why, as late as 1924, the period of engagement of a geisha with a guest was calculated by incense. A geisha was not just a prostitute but a trained professional—the word actually means "a skilled person"—who was taught from an early age to dance, sing, converse, and play games in a way that would be entertaining for male guests. The time they whiled away together was measured in sticks of incense, an entire evening generally calculated as the equivalent of four sticks. The "clock" that calculated the time passed was a rectangular container generally made of sugi wood, its upper surface perforated with round openings from which projected metal or bamboo holders for the incense sticks. Each stick holder was labeled with the name of one of the geishas. As each geisha became engaged with a guest, an incense stick was dropped into her holder and ignited. As each stick reached its end, it was removed and replaced with a fresh stick, and the burned remnant was labeled and set aside, for the purpose of tabulating the client's bill.

MOMENTS OF PERFECTION

The transcendent powers of scent are a reminder that our relationship to the material world is not merely purposeful, industrious, progressive. Fragrance is fleeting and elusive and enters us stealthily, at the edge of consciousness, transforming mood, unearthing long-forgotten memories, influencing us without our conscious assent. This transformative effect is what makes them so beneficial to our well-being. And the ephemerality and liminality of scent, bound up with love and memory and death, makes it a vehicle that transports us to states of heightened consciousness.

For me, to create with scent is to touch magic, in a process that eludes words. Trying to capture the experience verbally, I fear, ranks me with those who feel that by naming nature they own it—and threatens to exile me from what most excites me about working with aromatic materials. Rather, the perfumer is like the alchemist, who undertook to convert raw matter, through a series of transformations, into a perfect and purified form. The philosophy of alchemy expressed the conviction that the spark of divinity—the *quinta essentia*—could be discovered in matter. The ultimate goal was to reunite matter and spirit in a transformed state, a miraculous entity known as the elixir of life (also sometimes called the philosopher's stone). Following the dictum *solve et coagula* (dissolve and combine), the alchemist worked to transform body into spirit and spirit into body, to volatilize that which is fixed and to fix that which is volatile.

Beautiful smells minister to the human need for transformation and transcendence. Yet transcendence does not mean ignoring the mundane world; indeed it is rooted there, in our conscious attention to the details that surround us. Mindful of the texture of the very air we breathe in, we break through the glass walls of predictability, repetition, and pragmatism: we are inspired.

Almost every meditation practice focuses on breathing as a way of connecting with the here and now, of cultivating a sense of inner peace and participation in the beauty that's around us, all the time.

Consciously smelling is a form of breathing, accepting an invitation to an inspiring journey.

Consider what the painter Agnes Martin called "Moments of Perfection":

> I would like to consider further those moments in which we feel joy in living. To some, these moments are very clear and to others a vagueness that can only be described as below the level of consciousness. Whether conscious or unconscious, they do their work, and they are the incentive to life. A stockpile of these moments gives us an awareness of perfection in our minds and this awareness of perfection in our minds makes all the difference in what we do.
>
> Moments of perfection are indescribable but a few things can be said about them. At such times we are suddenly very happy and we wonder why life ever seemed troublesome. In an instant we can see the road ahead free from all difficulties and we think that we will never lose it again. All this and a great deal more in barely a moment, and then it is gone.

FRAGRANCE FAMILIES

Register and intensity aren't the only ways of organizing our thinking about scent. Nor are the origins of aromatic materials—leaf, spice, resin—the only other fruitful way of thinking about them; in fact, there are many ways. Linnaeus created the first recorded odor classification in 1752. Since then there have been many attempts to organize aromas into families of likeness. Scent families are a convenient way to keep track of the amazing array of natural materials. Thinking of them in groups makes it easier for you to compare and contrast them and to use your observations to penetrate more deeply into the nature of the essence. The greatest understanding of an aroma comes through recognizing minute differences between essences that smell similar but not the same. This simple list, then, is just a starting point for thinking in fragrance families. It is by no means exhaustive, and it is not cast in stone. As you become more familiar with scent, modify and elaborate and refine it in accordance with your own perceptions.

Spicy allspice, cardamom, cinnamon, clove, ginger,
   nutmeg, black pepper
Edible cognac, chocolate, coffee, tonka, vanilla, bitter
   almond, sarsaparilla
Herbal mint, basil, sage, lavender, oregano, rosemary,
   thyme, tarragon, basil
Citrus bergamot, grapefruit, lime, orange, lemon, litsea
   cubeba
Floral geranium, jasmine, orange flower, rose, tuberose,
   ylang-ylang
Resinous frankincense, myrrh, Peru balsam, benzoin,
   storax

Woody cedarwood, fir, pine, bois de rose, sandalwood, aloeswood, spruce

Earthy oakmoss, patchouli, vetiver, labdanum, tobacco

Animalic ambrette, costus, civet, musk, ambergris, onycha, hyraceum

Strewing sweet-smelling herbs and barks to perfume rooms is an ancient custom, but Anne of Austria, wife of Louis XIII, originated a new method (described in the 1690 *Pharmacopée Universelle* by Nicolas Lemery). She had small birds sculpted out of perfumed pastes and suspended from the ceiling. These were called *oiselets de Chypre*—*oiselets* meaning "little birds," and *Chypre* for "Cyprus," known as the place from which heavy perfumes came. When the "birds" were set on fire, they rose in the air.

I've come across a number of old recipes for "Cyprus powder" or "Chipre powder," most of them using an oakmoss base and featuring orange blossom, cloves, ambergris, orris, and roses. Here's one from Cooley's *The Toilet and Cosmetic Arts in Ancient and Modern Times* (1866):

Reindeer moss, carefully picked over, and then reduced to powder. It has a very agreeable smell, and being very retentive of odours, is much used as a basis for the finest scent powders, hair powders, skin (dusting) powders, sachets, etc. Ragged hoary evernia possesses nearly similar qualities, and is often substituted for reindeer moss. Oakmoss, being very retentive of odours, though in itself lacking the agreeable smell of the others, is also often used to make Cyprus powder.

RESINOUS AND WOODY ESSENCES IN FRAGRANCE

Frankincense and myrrh essential oils and benzoin absolute are all base notes with light intensity that blend seamlessly with almost every other essence. Frankincense has a soft, fresh, and balsamic aroma with hints of lemon and orange, and myrrh is warm and spicy, evolving into a sweet balsamic drydown, the last note you smell before the scent evaporates entirely. Benzoin (a resin from the bark of various Styrax trees) has a soft vanilla aroma. Unlike most base notes, these three essences can blend with milder notes without dominating them, sliding into the crevices and remaining in the background of a blend. Not only is aloeswood, or *oud*, a base note with high odor intensity, the most expensive essential oil in the world, but authentic, high-quality versions are extremely hard to find. A good *oud* should have a complex and intense barnyard smell—rotting, earthy, strange, impossible to describe. It careens through any blend, giving it a depth and an earthy complexity.

Besides the more exotic (and expensive) tree aromatics, there is also an enormous and appealing range to the scented resins and gums of more familiar evergreens. While fir absolute is a base note with a jammy, almost strawberry sweetness—like a walk in the forest—fir essential oil is a top note of dry wood with notes of fruity balsam, and it smells like a freshly cut Christmas tree. Many different types of cedar oil are used for fragrance. Virginia cedarwood, my favorite, smells like pencils—mild, slightly balsamic, sweet, and clean, evolving to a drier scent over time. As a top note with light odor intensity, it blends easily with almost any other top note. Atlas cedarwood is robust and intense, with slight camphorous backnotes. Bois de rose (rosewood) is, as its name suggests, a woody top note of low intensity, tempered with rose notes. It adds a rose note to the top of a formula and blends easily with almost any other oil.

In general the woods and resins are easy to work with. The muted ones are forgiving, with a soft complexity, and have the ability to meld into an existing blend and harmonize with the other players. Even fir absolute, which is not docile or retiring, blends with almost any other essence; you only have to beware that it not make a sweet perfume too sweet. The big exception, in almost every way, is *oud*. Given its outsize intensity and expense—good-quality *oud* starts at twenty thousand dollars per kilo—it should be placed like a diamond in an exquisite setting that will show it off.

In the Arab world, there is an old custom of perfuming drinking water with frankincense, although, as Aida S. Kanafani observes in *Aesthetics and Ritual in the United Arab Emirates*, the custom "is gradually disappearing."

As frankincense burns in the censer, the nozzle of the pot, which has been washed thoroughly, is placed on the incense burner. It takes approximately two minutes for the thick fumes of the incense to fill the pot and to leak from the spout. Water is immediately poured in the pot, which is covered until the smoke in the pot vanishes. This indicates that the smoke of the frankincense has been assimilated into the water. The flavor of the water remains fragrant for about three days, at which time the procedure is repeated.

RANKINCENSE PERFUME

This perfume highlights the contrast between the incense smell of fresh frankincense, the jammy sweet warmth of fir absolute, and the round greenness of lavender absolute.

See pages 69–70 and 72–73 for basic blending procedures.

SOLID PERFUME

8 milliliters jojoba oil

Heaping ½ teaspoon grated beeswax

6 drops frankincense essential oil

6 drops fir absolute

4 drops lavender absolute

6 drops frankincense CO_2

OIL-BASED PERFUME

10 milliliters fractionated coconut oil

6 drops frankincense essential oil

6 drops fir absolute

4 drops lavender absolute

6 drops frankincense CO_2

BODY OIL

20 milliliters jojoba oil (for body oil)

6 drops frankincense essential oil

6 drops fir absolute

4 drops lavender absolute

4 drops frankincense CO_2

ALCOHOL-BASED PERFUME

8 milliliters perfume alcohol

BASE NOTES

6 drops fir absolute

6 drops frankincense essential oil

MIDDLE NOTES

6 drops phenyl ethyl alcohol natural isolate

4 drops styrax essential oil

2 drops lavender absolute

TOP NOTES

3 drops tarragon essential oil

4 drops wild sweet orange essential oil

4 drops frankincense CO_2

The base of this perfume could not be simpler: even proportions of fir absolute and frankincense, with the fir sweetening the citrus and piney notes of frankincense. The styrax is a light and slightly floral resin that blends seamlessly with the phenyl ethyl alcohol, a natural isolate that smells like tea rose, and both are balanced by the sweet herbal yet floral freshness of lavender in bloom. The top brings forward green herbal notes, the sweet, anise-like scent of tarragon mixing with a much lighter CO_2 version of frankincense and the wild sweet orange providing a contrast of citrus that also highlights the citrus top notes of the frankincense CO_2.

INCENSE

When I burn incense, I find it's the odor of burning that I notice, rather than the subtleties of the scented material. My favorite way to experience incense is therefore not by burning it but by warming it. I use an electric incense heater, a small round censer with an electric heater in the base and a bowl on top. A dial controls the amount of heat, so that I can gently warm up and release the scent from resins and fragrant woods. I like to heat up a couple chunks of frankincense resin, letting the sweet balsamic aroma and fine but fragile top notes softly fill the room. My absolute favorite, hojari frankincense, delights the eye, nose, and hand: the resin looks like a cabochon jewel and leaves a subtle trail of fragrance on your fingers as you put it into the censer. Apart from frankincense, I tend to favor simple, discreet aromatics for incense, such as oakmoss, sandalwood, and benzoin.

Body incense has a long tradition in Japan, where it has been used as a scent, as part of a spiritual practice, and in the sleeves of geishas' kimonos to fragrance their bodies. There are a few Japanese companies that still offer this for sale.

Here are a couple of simple body incense blends you can make yourself:

CHYPRE-SPICE BODY INCENSE

> 2 tablespoons oakmoss powder
> 2 tablespoons sandalwood powder
> ½ to 1 teaspoon cassia powder

Place in a small bowl and stir to mix.

CHYPRE-RESIN BODY INCENSE

  2 tablespoons oakmoss powder

  1 tablespoon benzoin powder

  1 tablespoon frankincense powder

Place in a small bowl and stir to mix.

To use either blend: Place a tiny pinch of the body incense in the center of one palm and rub both palms together. Bring them to your nose and inhale. Rub some incense on your hair and arms.

BEARD OIL

You can create a simple soothing and nourishing beard oil by adding 20 drops of frankincense to 1 ounce of jojoba oil.

THE FRUITS OF TREES

Our most common methods of flavoring food with tree resins mimic their use as incense—using the wood directly in the cooking process to infuse the food with smoke or steam as in charcoal grilling, barbecuing, smoking, and cedarwood "planking." While some adventurous mixologists and chefs do use wood resins and essential oils directly in food and drink, few home cooks venture in this direction. But there is another group of tree-derived flavors with universal culinary application: the citruses. Their essential oils are mainly various terpenes—actually, the same family of aroma chemicals found in pine, cedar, and fir essential oils.

All citrus oils are top notes, with a light, bright, up-front sparkle of medium odor intensity. Having so much in common, they offer a great opportunity to train your nose and palate to perceive subtle differences in aroma and flavor. They all smell like their peels, which is the location of the essential oil. Lemon oil should smell "round," with a sweetness, not bitter. I prefer the smell and taste of pink grapefruit oil, with its uplifting and complex aroma, to white grapefruit. Orange smells are a world unto themselves and a study in subtle diversity. Sweet orange oil is sweet and fresh, whereas bitter orange is dry and delicate, with a floral backnote, and blood orange is rich and voluptuous, with notes of raspberry and strawberry. The lime oil that I prefer is from Mexico; it brings everything into focus with its sweet-tart, tangy scent. Bergamot has a sweet lemon-orange opening that evolves into a more floral scent, with a balsamic drydown.

Citrus oils deteriorate easily. They should be purchased in small quantities and stored in the refrigerator. After about half a year (or sooner if they begin to smell flat and lose their sparkle), they should be replaced.

An 1869 recipe for "Spruce Beer," from *Cooling Cups and Dainty Drinks*:

Sugar 1 lb.; essence of spruce, 1 oz.; boiling water, 1 gallon; add, when lukewarm, a tablespoonful of German yeast; ferment, and bottle like ginger beer.

DEANA SIDNEY'S FRANKINCENSE SHORTBREAD

Deana Sidney featured this unusual and delicious cookie recipe on *Lost Past Remembered*, her exceptionally interesting blog about historical food.

2 cups flour

½ cup sugar

1 cup (2 sticks) cold butter, cut into small pieces

4–7 drops frankincense (depending how strong you want the flavor)

1 or 2 drops lavender (depending how strong you want the flavor)

Preheat oven to 325 degrees.

Mix together the flour and sugar and cut in the butter with a fork, two knives, a pastry cutter, or your fingers; or pulse lightly in a food processor. Make sure that the butter remains in pieces the size of small peas in the mix—it should not be completely amalgamated. Add the frankincense and lavender and knead a few times, just until the mixture holds together.

Pat the dough into an ungreased 8-inch square or round baking pan, pressing it as flat as you can with your fingertips. (Or roll it out on a board and transfer it to the pan.) Pierce it with a fork, marking off small squares or rectangles or wedges.

Bake for about 30 minutes, checking it at 25 minutes. Remove from oven when very lightly browned and cut into pieces, following the tine marks. Transfer them to a rack to cool.

Store in an airtight tin.

A cachalot, or sperm whale, in a nineteenth-century engraving by J. W. Lowry.

CHAPTER FIVE

CURIOUS AND CURIOUSER

Ambergris

The cure for boredom is curiosity. There is no cure for curiosity.

—ATTRIBUTED TO DOROTHY PARKER

onder is a wonderful word, describing both our feeling of awe and delight at what we find strange and singular and new, and also the thing itself that arouses such a feeling in us. We seem hardwired for curiosity, and the objects and experiences that evoke it have a special place in our universe, neither deified nor disparaged but existing alongside us as a perpetually mysterious and magnetic Other.

The landscape of perfumery is rife with substances that trigger our sense of wonder, their origins so arcane as to seem mythic, with scents to match—funky, earthy, impossible to describe, at once attractive and repulsive. Perhaps the epitome of this world of wonders is ambergris, with its exotic origins in the digestive tract of sperm whales, and its surprisingly delicate, almost unearthly aroma. But the truth is, all the animal essences are rock stars. Visitors to my studio always want to smell them first; invariably they provoke a sense of wonder and excitement. As Steffen Arctander, author of the fragrance bible *Perfume and Flavor Materials of Natural Origin*, observes, "For some reason, it seems that whenever perfume materials are the subject of popular discussion, the few animal products, ambergris, castoreum, civet and musk, attract much more attention than do the materials of vegetable origin." As Arctander goes on to note, it was partly the provenance of these animal ingredients that gave them their aura of mystery and strangeness. In fact, the belief in their curative powers was a product of their exotic provenance as well as of their otherworldly aroma—not to mention their out-of-this-world costliness.

Fittingly, then, our path here will not be linear but meandering, driven by curiosity and passion and modeled on an artifact that both celebrated and was shaped by the universal love of oddity.

THE CABINET OF CURIOSITIES

The *Wunderkammern* ("chambers of wonders") were the forerunners of our natural history museums, and the first cabinets of curiosities were in fact rooms where people collected and displayed extraordinary, amazing, and bizarre objects from all over the world. They were not encyclopedic, and their goal was not to label, classify, organize, or explain. Indeed, anything ordinary or common was excluded in favor of the rare and strange, objects that seemed to represent nature's peak—or excess—of intensity or creative ferment. A premium was placed on whatever elicited a visceral "wow."

Striking visual demonstrations of the new valorization of curiosity could be seen in the cabinets of curiosities . . . that Renaissance princes collected and put on display at the courts. In these fabulous collections, the court projected an aura of the uncanny and the superhuman. Carved gems, watches, antiques, mummies, and mechanical contrivances were displayed side by side with fossils, shells, giants' teeth, unicorns' horns, and exotic specimens from the New World, making up an encyclopedia of the bizarre and the marvelous.

The items on display were the marvels of nature: the *mirabilia*. And although the word "curiosity" derives from Latin *cura*, "cure," and such oddities were often thought to have curative properties, a "curiosity" came to mean an unusual object that displayed the subtle and intricate workmanship of nature in its novelty, rarity, exoticism, or extravagance. Or even its tiny size. "Natural objects were often described as curious by virtue of their smallness, exquisiteness of workmanship being exhibited more strikingly in miniature." Curiosity was a virtue unto itself, and whatever sparked a resurgence of a childlike awe and delight was worthy of display.

By the sixteenth century, a cabinet of curiosities had come to mean the piece of furniture in which such a collection was displayed, a miniature museum whose many drawers, compartments, and shelves brimmed with natural wonders: bezoar stones, rubies, shells, skeletons of strange animals, bottles of unguents, a hunk of ambergris, specimens of taxidermy, musk pods. Out of their native context, silently juxtaposed with other curiosities, they seemed even more marvelous, a "more is more" feast for the senses of the natural world's infinite variety of wonders.

A modern version of the cabinet of curiosity is Pinterest, observes Benjamin Breen, in (fittingly) a blog post. There we collect images that catch our fancy, with no interpretation or order beyond a pleasing layering of images, textures, and colors, a privileging of "the eclectic and the exotic": a digital *Wunderkammer*, in other words.

The objects that made up a curiosity cabinet followed circu-
itous pathways (from Sri Lankan beaches and Amazonian
jungles, say, to Parisian salons), in the course of which they
lost their original contexts, names, meanings. Objects that
had once embodied human culture, like sculptures and coins,
became mere ephemerata. Natural treasures—corals, gems,
ambergris, bezoars—likewise functioned as mere "curiosities." Did that horn come from a unicorn or a narwhal? was
a question few early moderns ventured to ask, because the
items in curiosity cabinets did not invite speculation into
origins. They had no labels, after all. No narratives. No
"memories" as objects or images. So, too, with Pinterest and
its ilk.

It's telling about who we are and what moves us that the way we collect and display images and information so closely mirrors those
cabinets of old. "Perhaps the internet loves curiosity cabinets because it is, itself, a curiosity cabinet," Breen speculates.

But as any dog can tell you, in the hierarchy of the senses, smell
is hard to top for its ability to pique our curiosity. Coming to
us sometimes from far away, scents envelop us, touch us, unmedi-
ated, unannounced, unlabeled. The fact that they are universally
produced—what living thing does not have its own distinct smell?—does nothing to dispel the mystery. We
respond to them each time deeply and viscerally—not,
at first, intellectually. We wonder at what they are,
how they got there, what story they tell, what they
portend.

Part of the adventure of being a perfumer is getting to inhabit and explore an even more height-

ened world of olfactory curiosities than most. Here, for your delight and wonder, are a few of them I have been privileged to discover.

ROLLING AND MARKING

Out for a hike, has your dog ever dashed ahead and started rolling around in cow manure or roadkill, to your horror and your pet's evident delight? Dogs seem to delight in adorning themselves with repulsive odors as if with a spray of fine perfume. The phenomenon is called scent rolling, and Pat Goodmann, curator of Indiana's Wolf Park, has had ample opportunity to observe it. "Scent rolling is probably a way for wolves to bring information back to the pack," she explains. "When a wolf encounters a novel odor, it first sniffs and then rolls in it, getting the scent on its body, especially around the face and neck. Upon its return, the pack greets it and during the greeting investigates the scent thoroughly. At Wolf Park, we've observed several instances where one or more pack members have then followed the scent directly back to its origin." This scent-smearing ritual isn't limited to stinky odors. In her studies, Goodmann placed a variety of substances in the wolf enclosures and found that wolves did not discriminate against sweet-smelling scents. Besides rolling in eau de cat, elk, mouse, and hog, they also rolled in mint extract, Chanel No. 5, Halt! dog repellent, fish sandwich with tartar sauce, fly repellent, and Old Spice.

Mammals also perform an activity called scent marking where they anoint a member of their own species, another species, or an inanimate object with their personal scent. Mice and rats transfer scent via pads on the bottoms of their feet. Deer, following one another on a narrow trail, step directly where the animal in front of them has stepped. Their musk pads lie in the cleft of their hooves, and each step concentrates the odor mark left by other members of the group.

Scent marking is used primarily for social communication, and less often for purposes of defense against predators. But coyotes, foxes, and wolves, along with domestic cats and dogs, squirt urine to mark their turf, especially when they feel encroached upon. "This begins, of course, as simple elimination: the disposal of the kidneys' waste products after filtering the blood," writes Lyall Watson in *Jacobson's Organ*. "But on its way out, urine picks up an amazing range of perfumes provided by the renal tubes, the adrenal glands, the bladder, and the secretions of male accessory sex organs such as the coagulating and preputial glands. So by the time urine finds its way into the world, it is a very personal product, a mine of information."

The most prolific utilizer of scent marking in the animal kingdom is the male tiger, who scent-marks up to six hundred times as frequently as he merely urinates. In fact, the fluid he marks with is not ordinary urine but is much more intensely aromatic. Its smell is so potent that it has a special name in Sanskrit. Any guesses? *Vyagra*, derived from the verb root meaning "to smell." Clearly, someone at Pfizer, manufacturer of the best-selling Viagra, has a sense of humor—and a flair for natural history.

Humans, of course, communicate through scent, too. Using perfume to attract a lover or to signal our aesthetic taste, recognizing a parent or a partner by his or her distinctive scent, catching a whiff of fear, we are linked to the animals we love to find so strange.

GETTING SKUNKED

Skunk scent is in a league of its own, for both sheer aromatic intensity and repulsiveness. Many seemingly noxious animal secretions become prize olfactory ingredients when properly tinctured, but no matter how diluted, skunk remains too foul for use in perfume. Luckily, the normally docile skunk issues a public service announcement before spraying, in addition to the distinctive black-and-white coat it wears all the time. When it is riled, it draws attention to its anal glands by raising its rear and its tail, at the same time lowering its head. If this behavior fails to intimidate, the skunk lets loose its pungent secretions.

The secretion is stored in two sacs located just outside the skunk's anus, and the skunk can spray it as far as ten feet. The active ingredient, so to speak, is butyl mercaptan, which, like tear gas, burns the eyes and causes them to water and at high concentrations can induce vomiting. Even at very low concentrations—the equivalent of a single drop in a thousand gallons of water—humans can detect it and find it off-putting.

Animal trappers well know the uses of scent in animal communication and are expert at exploiting it, creating what are in effect perfumes from natural scent to lure wild game. A few decades ago, a seasoned professional trapper made plans to use scent to lure arctic foxes near the Saskatchewan–Manitoba border, with some unexpected consequences, which he recounted in amusing (at least in recollection) detail. (Bonus: Note the instance of scent rolling that the narrative involves.)

Chances were slim that I would get a chance to catch a white fox, but since I was doing a magazine article on Cree Indian trappers, and would be there anyway, I thought I'd order my favorite fox scent and have it handy in case I got a chance to use it. When the mailman delivered the scent I noticed that he was holding the small box at arm's length. As he handed it to me, I knew why, because I detected, though faintly, the repulsive odor of skunk. Upon examining the bottle I found it was all in one piece with the top sealed with melted wax. But for some reason, a tiny amount of this nose-tingling odor had leaked out.

I had to haul this bottle well over a thousand miles and the only place I could put it was in a small trunk with my camping gear. In an attempt to keep things under control, I dropped it into a small plastic bag, looped a rubber band around the top, and dropped it into another bag and secured this one the same way. When I finished the job, the bottle of scent was inside six plastic bags, and was about as tame as I could find a way to make it. When I arrived, though the temperature was near 40 below zero, and the trunk was unopened, there was a lingering odor. By now the bottle had frozen and broken, and while the liquid was still confined in the plastic bags, the wild odor had seeped out.

I now knew I had to get rid of the broken bottle of scent—and fast. I dropped the whole stinking mess into a paper bag, grabbed a snow shovel, went about a hundred yards from the cabin, dug a hole about 5 feet deep, and buried it. As I packed the snow over the top, I heaved a deep sigh of relief because my problem was solved and no one had detected me. This smug feeling didn't last long. Not more than thirty minutes later I heard a loud commotion outside the cabin, and when I looked out, there was a pack of sled dogs at the scent location rolling, fighting, howling, and digging in the snow. To top it off, when the dogs dug into the plastic bags a gentle breeze wafted this fragrance across the entire village.

FRAGRANT BUTTERFLIES

Even avid butterfly collectors are unaware that the males of many species can be quite fragrant, generating a scent with which they try to envelop the female during mating. Young butterflies are nearly odorless; it seems to require fully formed wings to diffuse the odoriferous secretion sufficiently to generate the characteristic perfume. The scent itself is not that of the flowers the butterfly feeds on, as might be expected, but a panoply of fragrances as astonishingly varied as the perfume counter at Macy's. An intrepid researcher in the 1920s attempted to catalog them, aided by the aromatic acuity of his two young sons:

Heliotrope
Rank, mouldy, cockroachlike
Syringa blossoms
Dried sweet grass
Jasmine
Freesia
Violet
Orris root
Chocolate candy
Musk
Mango flowers

Cinnamon flowers
Roses
Sweet peas
Nabisco biscuits
Carrot flowers
Sassafras
Rabbit hutch
Old cigar boxes
A sable fresh from the furrier
Strong honey or coarse brown sugar
Newly stirred earth in the spring
Unusually strong batlike odor

THE APOTHECARY'S
WUNDERKAMMER

The vast and varied store of ingredients that has supplied healers, magicians, cooks, and perfumers over the centuries is its own kind of *Wunderkammer*, and it has been home to some strange specimens indeed. None is stranger than the substance known as "mummy." A fifteenth-century medical handbook referred to mummy as a kind of "spice or confection," but it was supposed to be an exudation from the heads and spines of embalmed corpses. Powdered mummy was once so common that hundreds—or perhaps thousands—of antique ceramic jars made to contain it still exist in antique shops, museums, and private collections.

You'll be relieved to know that despite the "spice or confection" label, the demand for mummy was basically medicinal, not culinary. Imported from Egypt and the East, first-rate mummy was supposed to have a foul odor and a pitchlike consistency. As a medical handbook boldly instructs, "You should choose that which is shining, black, bad smelling, and firm." Mummy was considered to have "binding qualities" that would help stop bleeding. "To treat spitting of blood through the mouth because of a wound or a malady of the respiratory organs," said the guide, "make some pills with mummy, mastic powder, and water in which gum arabic has been dissolved and let the patient keep these pills under the tongue until they have melted."

ANIMAL ESSENCES

The "organ" where the perfumer keeps her materials is still a hive of wonders, especially the cubbyholes where the major animal essences lurk: ambergris from the sperm whale, musk from the tiny musk deer, civet from the wildcat of the same name, and castoreum from the beaver. Two additional animal essences are useful in perfume: hyraceum from the Cape hyrax, and onycha from the mollusk shell.

Prized throughout history, everything about these exotic, fabulously expensive essences is curious and magical. At full strength they are nauseating to most people, but upon dilution they reveal an incredible beauty and delicacy. Although they have been used in medicine as long as in perfume, it is in scent composition that their true sorcery becomes apparent. They have a remarkable effect on the other ingredients in a perfume, a little like how a pinch of salt can make a dish sing. They have the ability to enhance the longevity of a perfume, and to transform its texture, adding luminosity. Talk about alchemy!

The use of some of these ingredients in perfumery is also controversial, because in some cases the scent ingredients cannot be

obtained without killing the animal. In other cases harvesting the aromatics has historically involved cruel treatment of animals, although some progress is being made in developing sustainable and humane methods.

AMBERGRIS, THE KING
OF CURIOSITIES

Now this ambergris is a very curious substance . . . soft, waxy, and so highly fragrant and spicy, that it is largely used in perfumery, in pastiles, precious candles, hair-powders, and pomatum. The Turks use it in cooking, and also carry it to Mecca, for the same purpose that frankincense is carried to St. Peter's in Rome. Some wine merchants drop a few grains into claret, to flavor it.

Who would think, then, that such fine ladies and gentlemen should regale themselves with an essence found in the inglorious bowels of a sick whale! Yet so it is. . . .

I have forgotten to say that there were found in this ambergris, certain hard, round, bony plates, which at first Stubb thought might be sailors' trousers buttons; but it afterwards turned out that they were nothing more than pieces of small squid bones embalmed in that manner.

Now that the incorruption of this most fragrant ambergris should be found in the heart of such decay; is this nothing? Bethink thee of that saying of St. Paul in Corinthians, about corruption and incorruption; how that we are sown in dishonor, but raised in glory.

—HERMAN MELVILLE, *Moby-Dick: or, the Whale*

In 2006, Loralee and Leon Wright found the modern equivalent of Aladdin's lamp while fishing on the beach in Streaky Bay, Australia. They had parked next to what looked like a tree stump sticking out of the sand. Upon closer inspection Loralee decided that it was not a stump. Leon thought it seemed like some kind of cyst and suggested that they put it in the car and take it home, but apparently a thirty-two-pound cyst didn't fit in the category of things Loralee wanted in their vehicle, so they left without it. Over the next couple of weeks, though, they couldn't stop thinking about the mysterious object, and, their burgeoning curiosity getting the better of them, they went back and hauled it home. They called a marine biologist to come inspect their find, and he confirmed it as a chunk of ambergris, worth more than a quarter million dollars.

> The 1873 guide *Instructions and Cautions Respecting the Selection and Use of Perfumes, Cosmetics and Other Toilet Articles*, by Arnold James Cooley, comments on the centuries-old practice of creating scented pomanders, or *"pomambra"*:
>
> > Balls composed of ambergris, musk, and civet, beaten up with some of the stronger aromatics, as cloves, cinnamon, &c., and some excipient to give them form, were formerly used by embalmers to fill the vacant orbits of the eyes.

What is this mysterious substance, and what makes it so valuable? It must have been hard for the sailors who first found the soft lumps adrift on the water like baby Moses or the dry, porous lumps washed up on beaches and baked by the sun to imagine their source, let alone a use for them. Yet ambergris has been an article of trade in parts of Africa since about 1000 B.C. During the same period, the Chinese knew ambergris as *lung sien hiang*, or "dragon's-spittle fragrance," a reference to its supposed origins in the spittle of sea dragons sleeping on sea rocks and drooling into the ocean. It isn't surprising that these odd, unearthly masses that people found washed up on the beach gave rise to so many myths about where they came from, as their appearance gave so little clue to the true story.

A 1667 account listed eighteen theories on the origin of ambergris. One theorizer concluded that it was a rare and odorous kind of earth, because it was veined and marbled as earth sometimes is; moreover, it seemed likely that entire islands of the stuff must exist, of which the found pieces were mere fragments. Another theory held that ambergris was a gum exuded into the sea from the roots of certain trees growing along the coast. Ambergris even made its way into fiction; in *The Thousand and One Nights*, Sinbad tells how, shipwrecked on an island, he discovered a spring of crude ambergris that flowed like wax or gum to the sea, where it was swallowed by monsters of the deep. "The ambergris burned in the bellies of these whales and they vomited it. It then congealed on the surface of the sea, its color and quality were changed, and the waves cast it ashore, where it was collected and sold. The ambergris that did not flow to the sea congealed on the banks of the stream and perfumed the whole valley with a musk-like fragrance."

It wasn't until a sperm whale fishery had been established in New England that it was finally understood that the sperm whale did not swallow ambergris but produced it. In fact, Melville's Ishmael in *Moby-Dick* comes close to the truth of how the substance is formed. Ambergris is a waxlike pathological growth found in the stomach and intestines of about one in a hundred sperm whales, and also in the pygmy sperm whale. Its exact causes are still uncertain, but its growth may be spurred by the irritation created by indigestible elements in the foods eaten by the whales, such as the beaks of cuttlefish, which are almost invariably found in ambergris. Sometimes the whale will sicken and die before the condition can remedy itself, but usually the "raw" ambergris is expelled as a stinking fecal mass.

The various flights of fancy regarding ambergris were perhaps further inspired by the seemingly miraculous transformation of the material from this fresh state—fecal, soft, dense, and jet black on the outside and dark brown inside—to the aged white, porous, and faintly sweet-smelling chunks. Fresh black ambergris is not prized; considered low-quality, the gelatinous, soft, sticky mass has a strong manure note that can be described as smelling like a stable. But as it floats on the ocean for years and decades, a white coating forms on the outside (through a myriad of chemical processes including oxidation from the salt water), and the inside also becomes lighter in color as it dries and cures, with the fragrance growing lighter and more refined. The finest ambergris develops an incomparably lovely, sweet, musky odor that seems to combine perfume, the sea, and some primordial animal scent. No wonder Ishmael likens the miracle of its transformation to our own potential for transformation!

Ambergris was first introduced into medicine, cookery, and perfumery by the Arabs, who called it *canbar*. In Middle French this became *ambre gris*, or gray amber, to distinguish it from yellow amber, the petrified resin. Such distinctions were necessary because the resinous amber and ambergris were thought to have the same or similar origins. The ability of ambergris to stabilize or "fix" a perfume has dazzled perfumers since it was first discovered. Chemists have synthetically reproduced many of its scent molecules, marketed under various trade names, but still we lack complete understand-

ing, let alone replication, of all the scent nuances that make up natural ambergris.

Ambergris was initially prized most for its purported medicinal properties, as the preemptive weapon of choice against the foul miasmas that were thought to cause epidemics, especially the Plague. In the aftermath of the Black Death, the medical faculty of the University of Paris recommended carrying around sweet-smelling ingredients in "ambergris apples" (*pommes d'ambre*, the origin of the English word "pomander"). These were openwork metal balls that could be filled with various combinations of aromatics that varied according to recipe, availability, and budget. They were usually worn around the neck on a chain and so could accompany the wearer through the dangerous infested streets, an advantage over medical incense. The University of Paris "house blend" for pomanders called for storax, myrrh, aloeswood, ambergris, mace, and sandalwood. As is often the case, however, the highest-status version (and therefore the one presumed to be most effective) was also the most quietly expensive. The king and queen of France, according to their Paris doctors, were supposed to carry lumps of pure ambergris in their pomanders, which, ornately fashioned from silver and other precious metals and studded with gems, turned into a status symbol in themselves.

Only later did ambergris become as valued in perfumery as for its medicinal properties. For the chemist, ambergris remains one of the great mysteries of perfumery; a fixative of great value, it is long-lasting and mellowing. Used in small quantities, it creates an exalting and shimmering effect on the entire perfume. Sweet and dry, with stronger notes of wood, moss, and amber, it has only a slight animal aroma.

The German Georgius Everhardus Rumphius, one of the great

tropical naturalists of the seventeenth century, was known as the "Indian Pliny" because he was the author of *The Ambonese Curiosity Cabinet*—itself a true curiosity, brimming with fascinating details about the wonders of natural history he observed in a career with the Dutch East India Company, stationed on the island of Ambon in eastern Indonesia. His entry on ambergris is filled with charming lore, science, and firsthand experience, and Rumphius attempted to capture its fragrance by comparing it to that of onycha (*Unguis odoratus*), perhaps because that was the only other fragrance ingredient that came from the sea:

> Our people compare it to dried cowshit; and if it has some sand, small stones, or shells, then this does not make the ambar of a worse quality, but it lowers the price: If the seller does not permit you to break balls into pieces, though you would like to know if it contains sand, pebbles, and the beaks of Seacats, take a long needle which you have heated, thrust it into them, and you will become aware whether the needle encounters anything that is hard, whereafter you can talk about the price . . . [the same price] of ordinary gold. . . . One can be more certain, if one puts a piece on a hot iron or knife, because it should melt like wax and burn completely away or go entirely up in smoke, without leaving anything behind. One cannot describe the smell, which one must have experienced, because ambar has a particular smell, which cannot be counterfeited: we called it a Sea smell before, and the closest to it is the smell of the *Unguis odoratus*.

The first ambergris I owned was a tincture that was over a hundred years old, bought from a collector of whale antiquities. We were both bidding on the ambergris at an auction on eBay. As a then unseasoned bidder, I put in hundreds of dollars as my highest bid at a moment in the auction when the bottle of ambergris was only valued at thirty dollars. Of course, in the last five minutes the price skyrocketed way past the hundreds I had offered. This was back in the days when you could see the e-mail address of the person who'd won a particular auction, and I wasted no time telling the winner all about my work as a perfumer. I sent him a copy of *Essence and Alchemy*; I sent him perfumes; in short, I tried to seduce him into selling me his bottle of ambergris. He didn't, but we became fast friends, and since I had never smelled real ambergris, he kindly offered to put a syringe through the cork stopper of the bottle and suction up a few drops of the ambergris tincture to send to me. Fortunately for me, the cork disintegrated, and he sent me the many shreds to smell. I was in heaven. It was the most gorgeous smell I had ever encountered: shimmering, ambery, jewel-like, warm, luminous, and indescribable.

Years later he told me that he had decided to sell off his collection of whaling paraphernalia to a museum. His wife wanted the collection to be complete, including the ambergris. But, mercifully, he wanted me to have that bottle of ambergris, and I paid handsomely and gladly for it. I have never used it in a perfume but keep it to show to my students when I teach my perfume classes.

A company in New Zealand that has used me as a consultant carries many different varieties of ambergris, each, I discovered, with its own aromatic profile, which I have attempted to describe:

"marine"—fishy, not smooth, driftwood, beachy
"sweet marine silver"—beachy, but dirtier and sweeter than
    "marine" and less salty

"pure white"—seashore beachy, nothing ambery

"earthy woody"—earthy, woody, dirtlike, no marine notes

"antique"—strongest and most intense, sharp, some foulness

"white gold"—ambery, no marine, smooth and clean,
   beautiful

"sweet black"—my favorite, very animal and sweet, tarlike,
   smooth and slightly fishy, ambery

Ambergris was an ingredient not only for fragrance but for flavor. I have tasted ambergris tincture on chocolate and found it added a delicious warm ambery flavor. Robert May's seventeenth-century cookbook *The Accomplisht Cook* has many recipes that feature spices and other aromatics, including ambergris and musk and sometimes civet, steeped in rose water. And the nineteenth-century author Pierre Lacour writes about incorporating ambergris into drinks: "Ambergris is used as a perfume for liquors. It is never used alone, always being combined with other aromatics. The usual form of adding it to spirit, is to rub it well with sugar, which acts by minutely separating the particles of ambergris. Ambergris should be used in very small quantities, when used as a flavoring ingredient, as the odor would be easy of detection. In light-bodied liquors, one grain will often suffice."

Italian courtiers would sprinkle ground ambergris into tea, and a small piece of ambergris placed in the bottom of a cup would aromatize successive cups of coffee for two or three weeks. By 1644 chocolate became known in Italy as a medicine, with ambergris high on the list of flavoring agents; soon the court physician to Cosimo III de' Medici was experimenting with new flavor combinations, aromatizing the grand duke's chocolate with ambergris, musk, jasmine, and citron and lemon peel.

This apparently compelling culinary combination of ambergris and chocolate endures, with the famous gastronome Jean Anthelme Brillat-Savarin touting, via a recipe for "Chocolat Ambré," its almost magical healing powers:

All of those who have to work when they might be sleeping, men of wit who feel temporarily deprived of their intellectual powers, those who find the weather oppressive, time dragging, the atmosphere depressing; those who are tormented by some preoccupation which deprives them of the liberty of thought; let all such men imbibe a half-litre of chocolat ambré, using 60 to 72 grains of amber per half-kilo, and they will be amazed.

Sir Hugh Plat's 1609 *Delightes for the Ladies* gave the following suggestions for scenting gloves:

DIVERSE EXCELLENT SCENTS FOR GLOVES, WITH THEIR PROPORTIONS AND OTHER CIRCUMSTANCES, WITH THE MANNER OF PERFUMING

The Violet, the Orange, the Lemon duly proportioned with other scents, perform this well, so likewise of Labdanum, Storax, Beniamin [benzoin], etc. The manner is this. First lay your amber upon a few coals 'til it begin to crack like lime, then let it cool of itself, taking away the coals, then grind the same with some yellow ocher to perceive a right color for a glove: with this mixture wash over your glove with a little hairbrush upon a smooth stone in every seam and all over, then hang your gloves to dry upon a line, then with gum Dragagant dissolved in some rosewater, and ground with a little oil de Ben, or of sweet almonds, upon a stone, strike over your gloves in every place with the gum and oil so ground together, do this with a little sponge, but be sure the gloves be first thoroughly dry, and the color well rubbed and beaten out of the glove; then let them hang again till they be dry, which will be in a short time. Then if you will have your glove to lie smooth and fair in hue, go over it again with your sponge, and that mixture of gum and oil, and dry the glove yet once again. Then grind upon your stone two or three grains of good musk, with half a sponge full of rosewater, and with a very little piece of the sponge take up the composition by little and little, and so lay it upon your glove lying upon the stone. Pick and strain your gum Dragagant before you use it. Perfume but the one side of your glove at once, and then hang it up to dry, and then finish the other side. Ten grains of musk will give a sufficient perfume to eight pairs of gloves. Note also that this perfume is done upon the thin Lambs leather glove: and if you work upon a kid's skin or goat's skin, which is usual leather for rich perfumes, then you must add more quantity of the oil of Ben to your gum, and go over the glove twice therewith.

MUSK

The word "musk" is from the Sanskrit *mushkas*, which means "testicle." Musk is known to have been used in medicine and perfumery since 3500 B.C. It was apparently used for ritual purposes in ancient China and India, and the Carthaginians and Phoenicians were also acquainted with it. In the eighth to tenth centuries, musk was used as an antidote for poison (it was known as "snake terrifier"), for treating wounds, and for stopping bleeding, as well as a general restorer and kind of superdrug. By the tenth century A.D., the Arabs were using it in perfume, and they brought it to the Middle East. A Silk Road–like musk network tied Tibet to Muslim lands. "The scent's mystical properties were so highly valued in Persia that musk was mixed into the mortared walls of mosques, giving the temples a permanent breath of paradise. Musk was commonly mixed into medicines and candies, and was sometimes even eaten outright, one 'grain' at a time." It also perfumed not only bodies (both women's and men's) but fashion accessories and leather goods, particularly gloves and book bindings.

Musk deer weigh twenty-five pounds and stand only twenty inches high. They are solitary and shy, and do not grow antlers. Only the mature male produces musk, which is stored in a hairy pouch just the size of a golf ball, in front of the penis. It is composed of several layers of skin, with two openings immediately above the animal's urethra. To render musk for use in perfumery, the dried gland is chopped into small pieces and left in high-strength alcohol to mature for many months, or even better, for years. It possesses a sweet, generous, aromatic intensity and longevity, bringing an elegance and a radiance to any perfume composition.

Musk was—and is—so popular for fragrance because it can be used either on its own or as an incomparable fixative. Long before people were concerned about the sustainability of the species, the search for synthetic versions had begun, because of the staggering price and rarity of natural musk, but the complex odor profile and effect of musk cannot be duplicated in a lab. By the beginning of this century, however, musk deer populations had been hunted to precariously low levels, and the animal had disappeared altogether across large parts of its original Himalayan range. Already in 1973, the deer's plight was recognized by the Convention on International Trade in Endangered Species of Wild Flora and Fauna (CITES). The CITES agreement limits trapping of the deer, and most countries (Japan and France excepted) greatly restrict trade in natural musk.

Over the past fifty years, there have been efforts in China to raise the endangered Alpine musk deer (*Moschus sifanicus*) on farms. A 2011 study concluded that musk deer could be farmed sustainably through proper management and breeding.

A recipe for a scented ink from *Dick's Encyclopedia*, 1891:

The pure article can only be obtained from China. A good imitation may be made with ivory black, ground to an impalpable powder, made into a paste with weak gum-arabic water, perfumed with a few drops of essence of musk and half as much essence of ambergris and then formed into cakes.

CIVET

Give me an ounce of civet; good apothecary, sweeten my imagination.

—WILLIAM SHAKESPEARE, *King Lear*

Pure civet is a crude, buttery-yellow paste that turns darker with age. At full strength the tincture smells fecal and nauseating, but when diluted it has a radiant, velvety, floral scent. It gives great effects in perfumes, smoothing out rough patches, adding a sense of shimmer, diffusion, and warmth.

Civet is a catlike animal with a long tail and a long, pointed muzzle like that of an otter. Its various species are native to Africa and much of Asia. The paste is the potent secretion produced by the perineal glands of both male and female civets.

Ethiopia has been the principal supplier of civet for the perfume industry since ancient times; it has been farmed there since the Queen of Sheba gave some to King Solomon, or so the legend goes. As with musk and the other animalic essences, it was also used as medicine.

To Europeans civet was a "new" animal—they could find no record of it in ancient sources. They imported the paste initially as medicine, but before long imported the animal itself in large numbers from Africa and Asia. "Probably no foreign animal had as great an economic impact as the civet cat and its scent," writes Karl H. Dannenfeldt. "With the large numbers of civet cats in Europe, physicians and perfume manufacturers became less dependent on foreign civet, which was received only after a long sea voyage and with probable adulteration." By the sixteenth century, the scent was all the rage. Among the New Year's gifts Henry VIII is recorded as having presented to his courtiers in 1532, a "shevet" is included.

These days civet paste from farmed animals continues to be imported into Europe and the United States, virtually all of it from Ethiopia, traditionally shipped in the horns of zebu cattle. Each dried horn holds about five hundred grams of paste, the amount one civet can produce over a period of about four years. Chemists have developed synthetic versions like civetone, but they can't quite replicate the complexity and effects of natural civet.

tion in both Europe and North America. In North America, their near annihilation was thanks to the productivity of the Hudson Bay Company, which auctioned off 3 million beaver pelts between 1853 and 1877. The company also trafficked in the castor sacs, with a transaction recorded as early as 1688. Nowadays regions with large native beaver populations typically limit the damage caused by overpopulation of the species by having an open trapping season.

HYRACEUM (AFRICA STONE)

Hyraceum, also called "Africa stone," is the petrified and rocklike excreta of the rock hyrax, a medium-size mammal that looks like a big guinea pig and whose closest living relative, oddly, is the elephant. Besides having a positive transformative effect on every other essence in a perfume, hyraceum is perfect for lending a "dirty" note to perfumes that are too sweet or too fresh.

Rock hyraxes are native to Africa and the Middle East and live in colonies of ten to eighty individuals, all of whom have a habit of defecating and urinating in the same location. The resulting deposits, once they have petrified (a process that takes hundreds if not thousands of years), are a sought-after ingredient in both traditional South African medicine and perfumery. The "stones" are brownish and brittle; when they are broken open, they release a dark oil with an intense, complex, fermented scent, sort of a cross between civet and castoreum. Like the other animalic essences, hyraceum makes an excellent fixative in perfumery, and unlike some of the other animalic essences, it can be harvested at no harm to the species, which is thriving. The stones are still hand-gathered, as they have been since antiquity.

ONYCHA

Onycha is an aromatic essence used in incense since ancient times, derived from the shell of an edible marine mollusk found in abundance along the shores of the Red Sea. The part of the shell used for scent is the flap (or operculum, meaning "little lid") that closes the mouth of its outer shell; its shape, like that of a claw or fingernail, gave rise to the Latin name for the aromatic, *unguis* ("nail" or "hoof") *odoratus* ("fragrant"). The Chinese name for onycha, *chia hsiang*, likewise translates as "plate aromatic," from the shape of the operculum. In Sanskrit, notes James McHugh, an expert on the use of aromatics in South Asian religions, its name is "sweet hoof," and its use extended from the Mediterranean to China and Japan: "Indeed, it is probably the most ancient animal-derived aromatic to have an extensive global use, being mentioned in ancient Babylonian incense recipes." Its name carried erotic resonance in Sanskrit, McHugh goes on to say, both evoking and punning suggestively with the words for conch, prostitute, and fingernail—all synonyms that are used for it in recipes in *The Essence of Perfume*, an early Sanskrit text and "lending that text at times a quite suggestive tone . . . especially the case given that scratching with fingernails was an activity associated with sex in texts on erotics."

The Chinese mixed onycha with aloeswood and musk, and it is a primary ingredient in *ketoret*, the sacred incense described in Exodus. Like the other animal-derived aromatics, onycha was also used as medicine. Arabic medical books dating back to medieval times recommend inhaling the fumes of incense made with onycha as a cure for stomach pains, liver ailments, epilepsy, and irregular menses.

In perfume, onycha is typically used in the form of choya nakh, an intensely fragrant dark brown oil that smells like a campfire on the beach, with smoky, woody, seashore aromas and a touch of amber. It is created by roasting crushed onycha shells over a fire (typically built right on the sand). Once the fishy smell of the shells has volatilized, a smoky note emerges and remains. The charred shells are then combined with cedarwood essential oil and distilled.

Like the other animal essences, onycha has a smell that on its own is not entirely agreeable and that changes as the shells are roasted for choya nakh. Georgius Rumphius recorded his impressions of its olfactory metamorphosis in 1705, in *The Ambonese Curiosity Cabinet*:

> When broken into large chunks and laid on coals, it will first smell like roasted Shrimp, but then will immediately veer towards Amber, or, as Dioscorides would have it, to Castoreum; so that, if censed by itself is not very pleasant, but when mixed with other incense, the same gives, so to speak, a manly power, and durability; for since most incenses consist of woods, resins, and saps, that have a sweet, flowery or cloying odor, one should mix the Sea Nail among them, in order to make them strong and durable. One can therefore compare this Unguis with a Basse in Musick which, when heard alone has no comeliness, but which when mixed with other voices, makes for a sweet accord, and maintains the same.

The very qualities Rumphius describes make choya nakh a highly desirable ingredient in perfume.

In addition to using choya nakh, I also like to use the less familiar essence made from the unroasted onycha shells in creating perfume. It isn't easy to find commercially, but I've concocted my own, by taking the shells (which have very little aroma in their raw state), grinding them in a mortar and pestle, and leaving them to tincture in very high-proof alcohol. After several months they release a leathery and animal aroma somewhat akin to castoreum.

Fragrance expert Paolo Rovesti traveled around the world to learn about lost perfume traditions and the aromatics of vanished civilizations. In his book *In Search of Perfumes Lost*, he recounts this experience while traveling in Iran:

> At Schiraz . . . we met a merchant who used three different odours to give the caftans of his three sons a strong smell. Asked why, he replied: "Perfume is like an identity card. I can recognize my children by their smell whenever they go past my suk. They can't get lost among the many people of the bazaar and the caravanserai. I would find them again at once. At home they wear odourless caftans. But outside, with their nice smell they attract sympathy, which is very important to me."

THE HUMAN ANIMAL

The animal scents have a particular affinity for, and perform magic on, human skin. Being of a body, they speak to the unique odors of our own bodies—they remind us, in fact, that as animals we have our own distinctive scent.

Other creatures, of course, don't need to be reminded. Mosquitoes pick their victims through the amino acids and hormones we emit. As Darwin observed, apes respond with excitement to ovulating women, and male goats and bulls are more active around menstruating women. Scent may help explain why some people have such a hard time handling animals—nothing faulty in their approach but rather, as biologist D. Michael Stoddart puts it, "Humans are enclosed in a powerful olfactory envelope, explaining why some people have little trouble handling animals while others have much difficulty. The knack may be nothing learned but may be the result of a particular odorous complexion." And whether we know it or not, we take our cues from scent in how we respond to one another, a phenomenon the philosopher Michel Serres lyrically captures:

> Lingering near the surface of skins—veils, complex and subtle tissues—[is] this or that indefinable scent, belonging exclusively to her or to him and signifying each one to the other, in conscent. We do not love unless our senses of smell find themselves in improbable accord, a miracle of recognition between the invisible traces which scud over our naked skins, as air and clouds float above the ground. Until death there remains within us this spirit, and the chemical and mystical sense of the written and spoken word; as far as the nose is concerned, the emanations of whomever we have loved remain. It returns to haunt our skin, at dawn on certain mornings. Love perfumes our lives, aromas resurrecting encounters in all their splendor.
>
> We used to embalm the dead, so that the memory would evoke those that had been loved by our forebears.
>
> Life itself announces its presence from afar with these balmy emanations.

The idea that each of us exists inside our own idiosyncratic "olfactory envelope" asserts itself when different people try natural perfumes on their skin; responding to our biochemistry, they smell different on each of us, which is part of their allure. And inhaling the animal essences puts us back in touch with our animal nature, which expresses itself nowhere more keenly than in our sense of smell, which, while it is no longer as constantly needed for our daily survival—to identify food and foes and mates—still "works" as designed. These scents are exotic and "other," but at the same time they make us aware of the profound rootedness of our appetites and impulses *in* the body. They embody the lust to be fully, physically, sensorially alive in our one and only skin, the anatomy that is our destiny. They remind us that we humans are yet another curiosity, mortal and wonderful. Without words or thought, smelling these essences returns us to our animal nature. In them we experience nature's strangeness, yes, but also a sense of identity: we recognize our kinship with other creatures; we locate ourselves among nature's oddities.

ANIMALIC AROMATICS

All the animalic notes are base notes with the ability to transform and exalt other aromatics in a blend, making a perfume seamlessly beautiful with no rough edges, like the smell of your lover's body. This has less to do with their individual aroma than with their alchemical effect on the other essences in the mix. Individually and properly tinctured, their essences are sublime and otherworldly. The smell of aged and well-tinctured ambergris is smooth and pervasive, like a combination of old books, cathedrals, and the sea. Although ambergris (unusual among the animal essences) has low odor intensity, it exerts a "lifting" effect on the other essences in a perfume, creating an olfactory shimmer. I think of it as the aromatic equivalent of van Gogh's *Starry Night*. You can use ambergris the way a good cook adds a pinch of salt to a dish—everything is instantly improved.

Most of the animal essences have high odor intensity and, as noted, can be too much to take at full strength: onycha with its castoreum-civet scent; choya with its intense campfire smell; costus with an aroma like a cross between violets and wet dog (but in a good way!), with old precious wood and musky notes. Hyraceum, despite smelling like a combination of civet and musk, is useful for toning down perfumes that are too sweet, as it has no sweetness itself. Ambrette, from the hibiscus seed, is commonly considered to be the botanical version of musk. At once musky, smooth, and rich, it smells like brandy or overripe fruit. When used sparingly, it has the effect of making floral notes more floral.

TINCTURING AMBERGRIS

With its exceptional delicacy and transformative powers, ambergris is well worth obtaining. Unfortunately, most tinctured ambergris is too weak for its full exalting effect to be harnessed, so I prefer to make my own stronger tincture of about 10 percent strength. If you are curious, it is not difficult to obtain raw ambergris (see Sources) and to tincture it yourself.

There are several different colors of ambergris—white, gold, gray, and black—each with its own particular aroma, depending on how long it has been tossed on the ocean, how old it is, and its particular makeup. I have smelled many varieties, and my two favorites are sweet black (animal and sweet, tarlike and smooth) and antique gold (smooth, clean, and rich, with a beautiful amber quality). To make 1 ounce of ambergris tincture you need 3 grams of ambergris and 27 milliliters of perfume alcohol. I grind up my chunks of ambergris to powder in a mortar and pestle and put the powder in a beaker. I add the alcohol and swish it around to be sure I assimilate every last speck. Then I transfer the mixture to the top of a double boiler and place it over very, very low heat for a couple of hours, stirring it every 15 minutes. I cool the mixture and put it in a bottle, sealing it tightly. Then I let it stand for several months to mellow and age, shaking it and turning it upside down every couple of days, until the tincture is ready to use in perfume.

A recipe for "White Ambergriese Cakes" from *The Accomplisht Cook* by Robert May:

> Take the purest refined sugar that can be got, beat it and searse it; then have six new laid eggs, and beat them into a froth, take the froth as it riseth, and drop it into the sugar by little and little, grinding it still round in a marble mortar and pestle, till it be throughly moistened, and wrought thin enough to drop on plates; then put in some ambergriese, a little civet, and some anniseeds well picked, then take your pie plates, wipe them, butter them, and drop the stuff on them with a spoon in form of round cakes, put them into a very mild oven and when you see them be hard and rise a little, take them out and keep them for use.

ACCESSORY NOTES

As most of the animal essences, unlike ambergris, have high odor intensity, they provide a good occasion to talk about the role of accessory notes in scent composition. Accessory notes can be top, middle, or base notes, but on an odor-strength scale of 1 to 10, every one of them is a 9 or a 10 (or even a 10-plus!). One drop of any of them might be equal to multiple drops of a weaker ingredient. They are powerful, assertive, and idiosyncratic; some smell obnoxiously strong and take some getting used to. And there is no predicting how they will combine with the other elements of a blend. Their trace presence can transform a blend, giving it a unique cachet that verges on the alchemical. Yet if used in more than minute quantities, they will create a disaster.

Some common accessory notes include bitter almond, petitgrain, choya, peppermint, cèpes, litsea cubeba, clove, and cinnamon. They can bring out a nuance of another essence or reveal an entirely unsuspected aspect of it. To work with them is to be intensely in the presence of the mysterious and the magical. Used well, they add something definitive to a fragrance, give it originality, as a scarf or a belt can transform an outfit into a striking and unique fashion statement. Some Italian cooks use anchovies this way, to lend depth and pungency to a dish without dominating it—indeed, the unsuspecting may not even know that they are there.

Accessory notes can be a point of departure for a blend or a late addition to it, but however they are used, they require careful consideration of the other ingredients' character, intensity, and duration. They are my favorite notes to work with, and I will often build a perfume around one or two of them, highlighting their subtle tonalities and colorations. More than any other essences, they require experimentation and study to discover their possibilities. Spending time combining them with blander essences will yield countless ideas for the perfumer, shedding light on the architecture of sensuality.

BURYING

With all their trace molecules intact, natural essences are intrinsically complex. This is what makes composing with them so complicated and fascinating—and difficult. Sometimes a phenomenon in blending occurs that I call "burying," when an essence or a particular aromatic aspect of an essence is muted or eclipsed by another essence. Burying may be a good thing if it serves the overall design of the perfume—for example, when an essence with extremely high odor intensity finds a way to nestle into a formula without dominating. Choya nakh, for example (the essence made from onycha shells), might leave only a subtle feathering of its incomparable smoky note on the edge of a base chord when blended with other ingredients that can stand up to (and partially bury) it. But sometimes an expensive ingredient seems to have gone AWOL in a blend, because it has gotten buried. For example, boronia absolute, an incredibly rare and elegant (and ultra-expensive) ingredient, is of low odor intensity. It is very easy to bury underneath other florals, to the point that you'd never know it was there; this is a waste.

When thinking about composition, therefore, it's important to consider both the relative odor intensities of the ingredients and their odor profiles. For instance, trying to create a fresh, minty top note with peppermint and bergamot would be a mistake, because powerful peppermint will invariably bury weak bergamot. A much better choice would be to blend peppermint with an intense citrus like pink grapefruit, which is capable of standing up to it and maintaining an attractive relationship. Once again, essences are like people!

CHAMELEON PERFUME

Because of the difficulty of obtaining sustainable and ethically produced versions of some of the animal essences (and the difficulty and expense of obtaining many of them, period), I chose to use tobacco instead in this perfume. Its muskiness blends beautifully with the handsome floral of the orange flower absolute, and the pink grapefruit provides a sparkling opening.

See pages 69–70 and 72–73 for basic blending procedures.

SOLID PERFUME

8 milliliters jojoba oil

Heaping ½ teaspoon grated beeswax, melted

5 drops tobacco absolute

4 drops orange flower absolute

8 drops pink grapefruit essential oil

OIL-BASED PERFUME

10 milliliters fractionated coconut oil

5 drops tobacco absolute

4 drops orange flower absolute

6 drops pink grapefruit essential oil

BODY OIL

20 milliliters jojoba oil

5 drops tobacco absolute

4 drops orange flower absolute

8 drops pink grapefruit essential oil

ALCOHOL-BASED PERFUME

8 milliliters perfume alcohol

BASE NOTES

5 drops tobacco absolute

3 drops vanilla absolute

2 drops patchouli essential oil

MIDDLE NOTES

4 drops linalool natural isolate

3 drops rose absolute

4 drops orange flower absolute

TOP NOTES

4 drops Virginia cedarwood essential oil

4 drops pink grapefruit essential oil

2 drops bitter orange essential oil

A good tobacco absolute adds an animalic note to the base of the perfume. The other essences are highly mutable, and the experience of working with them will give you a good chance to see how a blend responds to minute additions of an animalic aromatic. Good-quality patchouli is a stunning essence for a perfume—like a fine aged cognac, it sweetens, lifts, and disappears. I added the vanilla for sweetness, not as a major player. Fine orange flower absolute is both very expensive and hard to find, but it rewards the perfumer with its suave masculine floral notes that are hard to place but beautifully combine with almost anything. The rose and linalool, a fresh light, floral wood, are here as simple middle notes. Together they allow the transformation from the animal notes, blending easily and beautifully. The top notes of citrus and light cedarwood are what I call filler notes: always welcome, easy to use, adding just the right amount of sparkle and dryness.

If you have the materials, you can also explore the effects of a couple of animal essences. After you have completed the perfume, divide it equally among four tiny bottles. Add a drop of ambergris to one bottle, half a drop of costus (dip a toothpick) to the second bottle, and a drop of hyraceum to the third bottle; leave the fourth bottle as is. Smell the four bottles and note the differences between them. Repeat after an hour, four hours, a day, and a week.

EARTHY AROMAS AND TASTES

As noted earlier, some plant-derived aromatics share the earthy and transformative qualities of the animal essences, whether used in fragrance or for flavor. Here are some suggestions on using them.

FOR FRAGRANCE: The full-bodied aromas of cèpes (mushrooms), coffee, vanilla, and chocolate can be used in new and exciting ways. Both coffee and chocolate are middle notes, while cèpes and vanilla are base notes. Cèpes and coffee have high odor intensity, while chocolate and vanilla have medium odor intensity. Cèpes absolute is the fungal, intense, somewhat "meaty" smell of wild mushrooms, and coffee CO_2 smells intensely of the bitter drink and, since it is a middle note, can be used to "dirty" up a floral. Since the aromas of these notes are so iconic, it is important to separate them from their connotations and clichéd uses. For example, blending with coffee essence requires you to let go of the "stimulating" association of the drink and to focus on its peculiarly roasted, earthy aroma. A good version of chocolate (sometimes called cocoa) absolute should smell mouthwatering, without a trace of bitterness or cloying sweetness. Like coffee, in very small doses it can be used to make a floral dirtier without registering as chocolate. The best-quality vanilla absolute, usually from Madagascar, is not merely a one-dimensional candy-sweet aroma. It is at once rich, balsamic, and treacly, but also woody, animalic, and spicy. The complexity of good vanilla is staggering, and this universally beloved scent can be used, in small doses, to enliven bitter blends—and, for that matter, any blend.

FOR FLAVOR: Since these earthy aromatics are so beloved and universal for flavor, as in perfume it is fun to think out of the box for them. One of my favorite uses for chocolate, for example, is to flavor beef. The two have a beautiful affinity that makes the beef more meaty and rich, like a subtle gravy. Add a drop or two of chocolate essential oil to a marinade for, say, flank steak, and you'll see what I mean. Both coffee and cèpes can be used, much like in perfume, to add an unexpected element of earthiness to asparagus, carrots, or potatoes. Try combining a drop of black pepper with vanilla and adding that to almost anything sweet or savory—it will make foods seem both familiar and new. I also love using vanilla absolute with spice essences to create a very special chai tea.

A recipe for "Wacaka des Indes," a chocolate blend using both ambergris and musk, from *The Druggist's General Receipt Book,* 1866:

Roasted cacao beans (chocolate) in powder 2 oz., sugar 6 oz., cinnamon ¼ oz., vanilla (powdered with part of the sugar) ½ dr., ambergris 3 gr., musk 1½ gr.

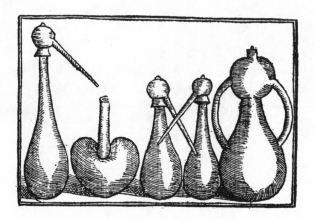

IASMINUM
PALLIDO
COERULEUM
PERSICUM
LATIFOLIUM.

A seventeenth-century engraving of a jasmine plant by Dutch botanist Abraham Munting.

CHAPTER SIX

SEDUCED BY BEAUTY

Jasmine

Beauty can be consoling, disturbing, sacred, profane; it can be exhilarating, appealing, inspiring, chilling. It can affect us in an unlimited variety of ways. Yet it is never viewed with indifference: beauty demands to be noticed; it speaks to us directly, like the voice of an intimate friend. If there are people who are indifferent to beauty, then it is surely because they do not perceive it.

—ROGER SCRUTON, *Beauty*

*I*n the world of perfume, flowers are a kind of shorthand for beauty. Their aroma is often called a perfume in itself, and rare is the great perfume that contains none of them. Take *Jasminum grandiflorum*, which is arguably the world's most important perfume material. Its small, white, waxy blossoms exhale a perfume so peculiar as to be incomparable. To walk past the flowering shrub in the evening is to be enveloped in the most glorious odor, which turns an ordinary street corner into a boudoir. The essence derived from the flower intensifies the experience. Jasmine is a deeply floral, warm, complex odor, with a peculiar honeylike, almost fruity sweetness that reveals a dirty undertone. Its almost cloying sweetness yields to a drier note as it evolves, but jasmine has considerable tenacity—staying power—and it retains its warmth and depth all the way to the drydown. "No perfume without jasmine" goes the adage that every perfumer knows, and to inhale it is to understand why.

At once voluptuous and delicate, earthy and ethereal, and elusive to those who would render these qualities immortal, flowers not only are beautiful but embody the paradoxes of Beauty that we embrace when we are drawn into her arms.

BEAUTY AND BEASTLINESS

*Is Jasmine then the mystical Morn—the centre, the Delphi,
the Omphalos of the floral world? Is it the point of departure,
the one unapproachable and indivisible unit of fragrance? Is
Jasmine the Isis of flowers, with veiled face and covered feet,
to be loved of all yet discovered by none? Beautiful Jasmine!
If it be so, the Rose ought to be dethroned and the Inimitable
enthroned in her stead; suppose we create a civil war among
the gardens and crown the Jasmine empress and queen of all.*
—CHARLES DICKENS, *Household Words*

"There is no beauty without imperfection," the saying goes, but it is too tame. As any aromaphile knows, there is no beauty without ugliness. Jasmine embodies the yin-yang duality—at once sultry and dirty, even fecal—that makes scent a catalyst for sexual attraction. Its narcotic lushness reminds us that beauty needs ugliness to exist at all. Pure gorgeousness would be bland and insipid without the foil of its counterpart; "the foul and the fragrant" are flip sides of a single experience, to borrow the title of Alain Corbin's great study of odor and hygiene in seventeenth- and eighteenth-century France.

Great perfumes capitalize on the foul-fragrant duality, which is often embodied in a single floral essence. In some, such as jasmine, tuberose, and orange flower, the magic (foul) ingredients also occur in human feces, including *indole* (with its hallmark fecal smell), *skatole* (from the Greek for "dung"), and *cresol* (with the coal-tar smell of its common source, creosote). While these scatological aroma chemicals are overpowering at full strength, they smell flowery at low concentrations, lending jasmine the putrid-sweet, sultry-intoxicating nuance that makes it such an aphrodisiac. Although many modern commercial perfumes depend on cheaper synthetic approximations, the great florals' complex fragrances, contradictory and elusive, are impossible to replicate synthetically. In the first place, it is hard to synthesize the specific mixture of several different molecules and their proportions as produced by the flower. That is why attempts to replicate jasmine essences result in an unpleasantly dominant note such as indole and demonstrate the limitations of the synthetics in general. Furthermore, nature, in the composition of its odor complexes, likes to include, in addition to the quantitatively predominant and identifiable components, minute amounts of materials that, by virtue of their intensity, play a decisive role in the character of the whole and its delicate "naturalness."

As isolated elements, the "foul" aromatics lose their magic, much as acting in a determinedly sexual manner often isn't sexy. But as an element of a natural essence, they ply the fine line between arousal and disgust, orchestrating a genuine eroticism. As in nature itself, complexity and context are the field conditions for awakening passion. The intensely earthy scents of the body that trigger libido are not, for example, erotic in themselves, any more than the blatant, unmodulated come-on of a "sexy" synthetic blend is. Scent can be sexual without being erotic. In our sexuality we are purely in the domain of nature; in our eroticism we are specifically human.

BOTTLING BEAUTY

The perfumer's pursuit of flowers, the most delicate and elusive of scented materials, has gone on for centuries. Some, like rose, can be captured through simple distillation, a process poetically described in this verse from *The Thousand and One Nights*:

> *My visit is shorter than a ghost's*
> *between Winter, it is, and Summer.*
> *Hasten to play with me, play with me;*
> *Time is a sword. . . .*
> *I wear my beauty and a crystal shift of dew.*
> *Men hurry me from my green to another crystal;*
> *My body turns to water, my heart is burned,*
> *My tears are collected*
> *And my flesh is torn.*
> *I feel the passion of fire,*
> *My flesh is fumed off,*
> *My spirit goes in vapour.*
>
> *The passionate*
> *breathe the musk of my cast garments with delight;*
> *My body goes from you but my soul remains.*

Other flowers, such as lily of the valley and sweet pea, still elude capture, leaving only the meager satisfaction of synthetic approximations.

As with many flowers, jasmine blossoms are too delicate to survive a steam distillation process. These days the aromatic jasmine absolute is produced through a solvent extraction process. But before that was invented, there was *enfleurage*, a process as intricate and sensual as the essence it yields. Enfleurage makes use of the fact that the volatile perfume materials of flowers are soluble in fat. Glass plates, each supported in a wooden frame, are coated on both sides with a suitable solid fat such as lard or tallow, which is prepared during the winter off-season by arduous melting, cleaning, and filtering until it is nearly odorless. Flower petals are laid on the plates, and the plates are piled on top of one another. As the petals emit their volatile elements, they are captured in the layers of fat above and below. When all the perfume of the petals has been absorbed by the fat, they are replaced by a fresh supply, and the process is repeated until the fat is saturated with scent. This fat, known as a pomade, is then dissolved in an alcohol-based solvent in order to obtain the aromatic essence.

It takes a little over a thousand pounds of flowers to produce a little less than one pound of jasmine absolute this way. The jasmine flowers release only a tiny fraction of their perfume at a time, but the fat captures a whole day's worth of fragrance from each blossom. By absorbing the flower's perfume over time, enfleurage captures more of its essence than could have been sniffed at any given moment of its life. A more apt metaphor for the quest for beauty—for obtaining the

essence of a thing that is somehow greater than the thing itself—
would be hard to imagine.

WABI-SABI

The Japanese design concept of wabi-sabi is based on appreciating
the transient beauty of the physical world. Wabi-sabi stems from,
even heightens and celebrates, impermanence: Nothing is perfect,
nothing is finished, and nothing lasts. It acknowledges that we and
everything around us are part of the natural cycle of growth, decay,
and death.

When I came upon this way of thinking, I felt I had discovered a
profound articulation of the unstated aesthetic that governs my ideas
about scent: the embrace of mutability and demise, the loss inherent
in the experience of beauty. Wabi-sabi elicits a yearning for something that defies definition, at the same time a sense of peace brought
on by the reaffirmation of our impermanence. And what better expression of our impermanence than the fleeting, volatile, evolving,
there-and-not-there experience of perfume?

It may seem antithetical to the lushness of the perfumer's materials to invoke a philosophy associated with simplicity and austerity,
but precisely because the materials are so sumptuous, the perfumer
risks creating a garish, overembellished, rococo edifice if she doesn't
have a disciplined creative process. An important principle of wabi-
sabi is restraint; nothing should be included that doesn't belong. Or,
to put it a different way, art can be better defined by what is left out
than by what is put in.

Beauty often thrives under physical or structural limitation. As
John O'Donohue notes in *Beauty: The Invisible Embrace*, traditional

cottages in the west of Ireland always had small windows, despite the beauty of the landscape, because the climate was rainy and cool, and small windows conserve heat. "Yet a small window exercised a discipline of proportion in relation to the external beauty," O'Donohue observes. "It never offered you the whole landscape: instead from every angle you look, it chose from the landscape a unique icon for your eyes. The grace of limit suggested more than your eyes could visually grasp."

To me this principle leads to perfumery at its most inspired. With so many essences to choose from, it's tempting to be seduced by all those beautiful smells, like glittering jewels on my perfumer's organ, with hundreds of bottles of essences. But there is a no-frills economy to wabi-sabi: No element should be surplus. "The main strategy of this intelligence is economy of means. Pare down to the essence, but don't remove the poetry. Keep things clean and unencumbered, but don't sterilize." This means that no essence goes into a perfume formula that does not have its place—rigorous editing eliminates any ingredient that isn't used to advantage or is merely decorative. A successful perfume is a smell that never existed before, yet it must feel whole, its elements balanced in a way that appears natural and unforced.

Another tenet of wabi-sabi is that the form of an object should emerge from the physical properties of the materials with which it is composed. In my world of natural perfume, I create within the boundaries of the botanical world. Wabi-sabi disregards conventional views of beauty, looking for it instead in small, barely perceptible details. My own aesthetic takes root in the irreducible nuances of the materials I work with; the idiosyncrasies of their unique footprints inspire unconventional combinations.

When I first begin creating a new perfume, I usually have a lot of ideas that I have to try. But often they fail to satisfy. Finally, when I am on the verge of defeat, something comes from the rubble. After many blind alleys, I find where to begin. One essence calls for the next, and at last I am capable of listening for notes that bear the stamp of inevitability, as if I couldn't have chosen any other way. I am on the lookout for what is aesthetically necessary—technically, intellectually, and intuitively—not aiming for what is "beautiful."

Boro are Japanese textiles sewn from nineteenth- and early-twentieth-century rags and patches of indigo-dyed cotton. The word *boro* means "ragged," and the range of patches on each piece is a kind of tour of antique hand-woven fabrics. Before the twentieth century, large parts of Japan were so poor that clothing and bedding had to be mended repeatedly as they were passed down from one generation to the next. The unplanned arrangement of patches and mending exemplifies wabi-sabi aesthetics that prize devolution, simplicity, incompleteness, and the deep attachment to meaningful and singular objects in one's life. Paradoxically, *boro* embody the wealth one feels when one knows what is really important and values it.

THE IDEA OF THE BOUQUET

Beauty is a quality that we feel rather than understand, a wave that overcomes us. The experience precedes language and defies piecemeal analysis. We can speculate endlessly about what makes something—or someone—beautiful, but we cannot quantify it. All the logic in the world cannot make something beautiful, and all the logic in the world cannot comprehend why something beautiful is so. Beauty does not reside in parts; it is a product of the relationship among parts, a relationship that is itself nongeometric, askew. It is beauty's holistic character that is responsible for its effect, a body blow that can literally take our breath away.

The experience of smelling a beautiful essential oil is like this, too. Although the different molecules that constitute it could be isolated and reproduced, their separate aromas would never equal the experience of it whole. And the perfumer seeking to create a perfume is after the same effect, feeling for an underlying connection that will bring disparate scents together into a whole that erases the lines between the parts without eclipsing any of them.

Easy to say, difficult to achieve! Just as when you attempt to furnish and decorate a room, making a series of strong choices does not suffice. We pick out a color we like to paint the walls, we buy a couple of chairs we think are of a nice design, a mirror that has an interesting shape, and so on. It may be that each of these decisions, considered in isolation, is pleasing. But they don't necessarily work well in unison. The principles of creating an ensemble are sometimes counterintuitive. Rather than buying curtains that are in themselves attractive, we might be better off buying some that are neutral and unobtrusive—that is, if we don't want to draw too much attention to them, if we want them to function primarily as a backdrop to the room. The curtains are there for the sake of the room; the room isn't an excuse for hanging curtains. The same goes for perfume: Each decision has to work in service to the effect of the whole.

When—against all odds—many elements come together to create a seamless and interesting new fragrance, I call the phenomenon "locking." I can think of no better metaphor for this process than Michel Serres's bouquet:

It is a wise and true language that calls the exhalation of the fragrance a bouquet. A bouquet is not just a mass of flowers, a simple multiplicity, but a bundle tied together, held by string or thread or the neck of a vase. Each flower adds its color and shape, spreads and diffuses its perfume, but each one vies with the others; bouquet expresses their intersection. If you pull towards you the knot, ribbon or neck, the precise place where a confusion of multiple cascades is formed, all the stems and petals will come together, the whole state of things is revived in your memory. No single component can be identified separately from the resultant. A bouquet forms a fragment of memory because of the impossibility of analyzing mingled bodies: either it has integrity, or does not. A singularity reappears around the intricate intersection.

Bouquet expresses a product, an intersection that defies analysis.

The seductiveness of many aromatics can make you wish that you could simply mix them together in a beaker and call it a perfume. Alas, just as mixing all the beautiful watercolors creates mud, a perfume blended without balance and structure is a disaster. Essences, like people, can fight, get along, bring out the best in each other, overwhelm each other, or come together and create something strange and new. An ancient Indian treatise held that a perfume should be like a well-run kingdom, "with the correct balance of allies (mild materials), neutrals, and enemies (pungent materials)."

The sweet, the foul, the spicy, and the putrid—I find them all alluring. I love the way they smell and the way they look, some like liquid rubies or emeralds in the light, some thick and pasty, others light and thin. I find them neither beautiful nor ugly in themselves. Instead, like colors, they are interesting to me for their textures, their aromatic shapes, their layers of complexity. A painter would not refuse to use yellow because yellow is an ugly color; she would simply use a particular shade of yellow where it was needed. "The different marks I employ must be balanced in such a way that they do not destroy one another," Henri Matisse observes in his "Notes of a Painter," explaining the role of color in his paintings. Beauty derives from context. The medieval theologian and philosopher Alexander of Hales noted that the universe is a whole that is meant to be appreciated as such, the existence of shadows only making the light shine more brightly and the ugly becoming beautiful by virtue of its place in the order of things.

A TENDER SORROW

I can barely conceive of a type of beauty in which there is no melancholy.

—CHARLES BAUDELAIRE, *My Heart Laid Bare*

Beauty is impermanent: it visits but does not stay forever. It envelops us in delight, like a vapor, and then, like a vapor, it evaporates.

In wabi-sabi, the effects of time are considered expressive and attractive. It treasures the passing of time and the patina that impermanence leaves in its wake. "There's an aching poetry in things that carry this patina, and it transcends the Japanese," writes environmentalist Robyn Griggs Lawrence. "We Americans are ineffably drawn to old European towns with their crooked cobblestone streets and chipping plaster, to places battle scarred with history much deeper than our own. We seek sabi in antiques and even try to manufacture it in distressed furnishings. True sabi cannot be acquired, however. It is a gift of time."

Sensing the beauty of the aging process allows one a more intimate and richer experience, as designer Andrew Juniper elucidates:

> The physical decay or natural wear and tear of the materials used does not in the least detract from the visual appeal, rather it adds to it. It is the changes of texture and color that provide the space for the imagination to enter and become more involved with the devolution of the piece. Whereas modern design often uses inorganic materials to defy the natural aging effects of time, wabi sabi embraces them and seeks to use this transformation as an integral part of the whole. This is not limited to the process of decay, but can also be found at the moment of inception, when life is taking its first fragile steps toward becoming.

Perfume, in all its manifestations, is a testament to the passage of time—depending on the passage of time for its effects. Time passes as fragrance evolves on the skin, telling a story, beginning to end, before it disappears entirely, leaving only a sweet memory. Before that story can even begin, the ingredients must marry in the bottle, like a fine wine, sitting in a dark cupboard as they acquire smoothness. During that time I test the fragrance on my skin and may make minor adjustments, until I think the perfume has ripened. Earlier still, the essences themselves must ripen, some in the bottle, some in their raw state, blossoming or fruiting on the vine or crystallizing, petrifying, tossing on ocean waves or drying on beaches over decades, even centuries. Among my perfume ingredients is an extensive collection of century-old essences, many of them so extraordinary as to defy description. Their aromatic molecules have merged and aged like those of a fine cognac.

In its acknowledgment of transience, wabi-sabi induces a serene, transcendental state of mind in the beholder. Andrew Juniper explains:

> Everything in the universe is in flux, coming from or returning to nothing. Wabi-sabi art is able to embody and suggest this essential truism of impermanence. Experiencing wabi-sabi expressions can engender a peaceful contemplation of the transience of all things. By appreciating this transience a new and more holistic perspective can be brought to bear on our lives.

This serenity is not the product of a denial of loss but rather takes solace in it. *Mono-no-aware* is an ancient principle of Japanese aesthetics that means "seeking solace in sadness." It "refers to the transient emotional elements of a work of art, as opposed to the ideal or objective elements," writes Boyé Lafayette De Mente in *Elements of Japanese Design*. "In effect, it means giving oneself up to tender, sorrowful contemplation of a thing or scene that is the opposite of sunny, happy, and bright. For example, the feeling evoked by the sight of falling cherry blossoms is *mono-no-aware*."

A formula for an incense known as "Bare Boughs," used in the incense "game" *kōdō* around 1400, featured aloeswood, musk, cloves, sandalwood, and onycha; it was intended to symbolize the loneliness of early winter. *Mono-no-aware* holds at its center the idea that the only constant is change. At a deeper level, it is an understanding of the exquisite perishability of beauty and human happiness. The pathos of the fragile and fleeting reality of life fills us with feelings of tenderness and sorrowful contemplation. As Donald Keene observes, the Japanese appreciation for mutability explains not only their celebration of the brief appearance of cherry blossoms, which fall three days after flowering, but also their choice of using wood for building temples, in preference to brick or stone.

Fragrance, striking us as simultaneously timeless (in its evocation of memory) and evanescent (in its fleeting beauty), gives us the opportunity to marvel at our precious life and the magnificence of nature even as the experience is tinged with sadness. Whether it lasts five minutes or for hours, it is not an endurance race against time; rather, it is poetry. And while it lasts, it invites the imagination to embark on a dynamic reverie, creating an internal unfolding— a kind of dream composed of successive images, thoughts, and feelings—that parallels the unfolding of the perfume itself.

In its inextricable connection to the passage of time, music, even more than painting, is a metaphor for how perfume is created and experienced. Is it mere coincidence that individual essences are called "notes" and are blended together to form "chords"? Or that the unit where I keep my materials and compose my perfumes is called an organ, its semicircular stepped shelves lined with a vast array of essences, ranged by register? Sitting at the organ, the perfumer can construct fragrance creations in much the same way that a musician chooses musical notes and composes chords. The musical scale is an apt analogy for the perfumer's palette (a painting metaphor intrudes!) precisely because its tones do not all fit together in easy consonance but are full of the potential for discord, with their idiosyncratic traits and competing intensities. Musical concepts like tone, vibration, and harmony resonate in perfumery as well, where the relationship between essences structures a blend just as musical structure depends largely on the relationship between tones. Each element, whether it plays a major or minor role in the final blend, modifies the character of the others. A fragrance of perfect balance and harmony—rich and mellow, smooth and perfectly toned—is called "round."

Music also captures the way scent is experienced—not all at once but unfolding over time—a quality that in perfume is referred to as "duration." Perfume can be "listened to" as an evolving form that moves through aromas, from sharp and pointed top notes to its well-developed, expansive middle notes, and finally the deep, dark, heavy base notes that bring it to a close. In their unfolding lies the unparalleled power of these arts over memory and emotion. Music and scent can calm us or arouse our passions—and, in our ecstasies, exalt us. They seize us, they transport us to the highest realms, feeding a desire for intoxication. They alter our consciousness in a way that static experiences or symbolic systems like language cannot, nor can their most transcendent effects be fully expressed by language. They are ineffable.

BEAUTY DOES AS BEAUTY IS

The soul is weighed in the balance by what delights her. Delight or enjoyment sets the soul in her ordered place. Where the delight is, there is the treasure.

—SAINT AUGUSTINE

What is beauty for? The beauty of beauty is that it is not *for* anything—it doesn't stand for something else, it doesn't have to *do* something, it only needs to be.

FRAGRANT

Beauty touches and renews our hope when it takes us out of the grid of ordinary time and brings us to another place, a place where history ceases and the weight of memory relents, a place ever ancient and ever new. . . .

No-one is immune to beauty. Regardless of background, burdens or limitations, when we find ourselves in a place of great beauty, clarity, recognition and excitement awaken in us. It is never a neutral experience. . . . To behold beauty dignifies your life; it heals you and calls you out beyond the smallness of your own self-limitation to experience new horizons. To experience beauty is to have your life enlarged.

Beauty's lack of purpose hardly means that it is not essential: beauty brings about a morally valuable state in the mind of the beholder. When we experience beauty, it lifts us up and returns us to our spiritual home. Beauty sustains an inner life.

I find it peculiar that some people think our relationship to nature should be eternally purposeful and industrious, always seeking greater use or broader knowledge. This dreadfully serious and puritanical approach is incompatible with the sensuality of scent. I create perfume, and people wear it, because beauty is a vacation from reality. It is a place—an ideal place—that you can visit without traveling. It is restorative, and it makes you feel good. A personal adornment like wearing jewelry, it has no practical purpose whatsoever. It simply allows us to inhale bliss.

The "industry" model also colors attitudes about what aromatic materials are acceptable for use in perfume. In a recent interview, Francis Kurkdjian, a formally trained French perfumer who has created scents for Jean Paul Gaultier, Yves Saint Laurent, and Guerlain and has an eponymous line of fragrances as well, likened reliance on natural aromatics to trying to construct "a modern city without steel and glass" and ending up with "just huts" as a result. "Synthetic notes are the backbone of the structure, the longer-lasting notes, or the steel beams in the building you're making," he said. "If you use only natural products, in two hours they will die on your skin and you will have no aura, no power." I have heard this argument a number of times, but I cannot for the life of me make sense of it. Why liken perfumery to construction? Buildings are required to serve many functions, perfumes are not. Which is not to say that they lack sophistication or ambition. There are many magnificent examples of edifices made without steel, from Gothic cathedrals to Arts and Crafts homes like my own. And every successful work of art, "impractical" as it may be—from music to sculpture to poetry—has structure.

But behind this clichéd argument against natural aromatics lies the real objection: Using natural materials violates the way the mass-market fragrance industry allocates the costs involved in making perfume. A former department-store CEO offered the website DailyFinance a rare glimpse into where the money actually goes, breaking down the profit-and-loss assumptions on which a typical 3.5-ounce bottle of a "celebrity" fragrance retailing for $100 is brought to market. Notably, having worked in an industry that guards such information as if it were Colonel Sanders's original fried-chicken recipe, the former executive remained anonymous in the article:

Bottle $6
Packaging $4

Marketing $8

Sales commission (for people pushing the fragrance at the department store) $6

Licensing fee (for a celebrity fragrance) $4

Manufacturer's overhead $15

Manufacturer's profit: $15

Retailer's overhead $25

Retailer's profit $15

"Juice" (the aromatic materials) $2

As the author of the article pointed out, "Despite all the flamboyant marketing-speak behind prestige fragrances—all that talk of floral formulas and gourmand notes—the value of the actual liquid is roughly equivalent to a large cup of regular coffee. Yep, not even a cappuccino." With this kind of formula—and the high overheads, high marketing and packaging costs, and high profits it is expected to fulfill—it's easy to see why "scaled-up" fragrances can't be made exclusively from natural ingredients, and why they tend to lack genuine luxuriousness, despite the fancy bottles and lush ads.

For me perfumery is an art form, and it all starts with the materials. I feel about aromatic materials the way Bob Dylan felt about songs: "Songs, to me, were more important than just light entertainment. They were my preceptor and guide into some altered consciousness of reality, some different republic, some weird republic." And the words of the painter Agnes Martin are a beacon and a reminder to me about why and how I do what I do:

The function of art work is the stimulation of sensibilities, the renewal of memories of moments of perfection. There is only one way in which artists can serve this function of art. There is only one way in which successful works of art can be made. To make works of art that stimulate sensibilities and renew moments of perfection an artist must recognize the works that illustrate his own moments of perfection. Perfection, of course, cannot be represented. The slightest indication of it is eagerly grasped by observers.

There is reciprocity between me and the person who wears my perfumes that is not just commerce: beauty is a happiness that we are sharing. Perfume is quite literally, for me, a message in a bottle. I am hopeful that what I experience when I make it is then experienced by the person who wears it, creating a kind of intimacy between us.

Smelling takes place in your nose but also in your mind and in your memory. A beautiful fragrance, like any fragrance, is ephemeral, but it lasts longest in your imagination, untethered at last from the things of this world. Again, Agnes Martin:

When I think of art I think of beauty. Beauty is the mystery of life. It is not just in the eye. It is in the mind. It is our positive response to life. We see everything in its perfection. We say a newborn baby is beautiful and when we enter the forest we do not see the falling trees and rotting leaves. We see the perfection and we are inspired. . . . The subject matter of art work is this response to perfection that we make the beauty in our minds. Some art works protest its absence but all art extols beauty.

Instructions for making rose "beads" or "pearls" from an 1873 "receipt" book:

These are commonly made of the *petals* of *red roses*, by beating them in an *iron mortar*, for some hours, until they form a smooth, black, stiff paste, which is then rolled or molded into "beads," small "balls," or "medallions," and dried in the air. To facilitate the process, a small portion of the petals are sometimes more or less air-dried. The addition of a few drops of *otto of roses* [rose oil] toward the end of the process improves them. . . . When quite dry they are very hard and fragrant, and take a fine polish. Sometimes they are turned in a lathe. Kept in dressing-cases, work boxes, drawers, wardrobes etc. to scent them.

A WABI-SABI WAY OF LIFE

Happy is he whose craft is that of a perfume-maker.
—BABYLONIAN TALMUD, KIDDUSHIN 82B

Oscar Wilde believed that everything we do, wear, and surround ourselves with is part of an overall enterprise of beauty: "In the question of decoration the first necessity is that any system of art should bear the impress of a distinct individuality; it is difficult to lay down rules as to the decoration of dwellings because every home should wear an individual air in all its furnishings and decorations." Wilde's ideas about self-expression and aesthetic choices were a precursor to what is now called "lifestyle": he lived by his own tenets and encouraged others to make informed and creative decisions about the decoration of their homes and the adornment of their bodies.

Taken to heart, Wilde's precepts influence not only your choice of shoes, furniture, and shrubbery, but also the design of, say, your résumé or your website. Everything you put into the world not only can be pleasing to the eye but can consciously convey your meanings and values. In fact, it *will* convey meanings and values, and it is best if they are not someone else's by default but rather those you practice and hold dear.

In the beginning I thought of myself as an artist who created beauty and then—on the downside—had to find a way to manage my creations' lives in the world. In other words, I had to develop a business. But over time I came to see the two enterprises as inextricably intertwined. I find the shape and evolution of the business I have created no less an expression of my fundamental aesthetic than the many perfumes I have made, and it is guided by the same principles. I run Aftelier Perfumes in a wabi-sabi way: I maintain a size and shape appropriate to what serves my own needs and the quality of the perfumes, which is paramount to me; I neither add nor do anything extra. Most businesses focus on growth—how they can sell more and expand their customer base. My focus has been on never allowing my business to become so big that I can't do everything myself. I love creating a handmade product that I see through from start to finish.

I believe that the world as it is now, troubled and unstable as it is, also allows many people a greater freedom to make just these kinds of choices. If perfume is intimately linked to time, I am lucky to have become a perfumer when I did. I would not have been able to establish a business of this scale before the Internet, which enabled me to avoid courting the big department stores and bending to their ways of doing things (which I knew from early experience was antithetical to maintaining an artisanal enterprise). Running a business like this is itself a great luxury; it allows for almost complete aesthetic freedom and requires very little compromise. Of course, the enterprise is labor-intensive, but the deep satisfaction of making something that is beautiful and doing it in a thoughtful, artistic way fills me with joy. I wouldn't have it any other way.

Long ago, before there were perfume houses and multinational conglomerates that made perfume, there were little perfumeries and apothecaries; now, thanks to the Internet, it has become possible to return to those roots in a sense, spawning a surge of artisan and niche perfumers. This revival has created a parallel revival in connoisseurship, as small-batch makers connect directly with increasingly sophisticated customers. In the world of scent, a busy colony of perfumistas has come into being, who are spectacularly knowledgeable about fragrance and passionate about indie perfumes. Many of them post about their finds and share decants (small samples) with discrimination, enthusiasm, and love. An erudite, sophisticated cadre of independent perfume bloggers share their passion and deep understanding of the aesthetics of artisanal and niche perfumery. In their own way, these consumers are manifesting their sensibility in the world, helping to support a welcome variety of perfume creativity at several levels. Beyond my own hands-on artisanal level of perfumery, there are many larger indie perfumers whose creative vision is directly borne out in their perfume lines, and many niche fragrances introduced by larger brands to showcase the creative talent of their perfumers.

The development of a personal aesthetic is not the exclusive province of the creative among us. It is one of the great joys of being human, and it is open to anyone. Indeed, everyone has an aesthetic, although in those who do not consciously develop theirs, it is apt to remain a stunted thing or, worse, a secondhand one, dictated by the pages of the fashion magazines and other received ideas about what beauty is. Conversely, inspired by what we surround ourselves with and immerse ourselves in, it can become something original and sublime.

For better and worse, our aesthetic sense is extraordinarily suggestible. Consider this experiment, which was performed in front of a live audience in 1899:

The following scheme for the production of a hallucination of smell may be worth recording. I had prepared a bottle filled with distilled water carefully wrapped in cotton and packed in a box. After some other experiments I stated that I wished to see how rapidly an odor would be diffused through the air, and requested that as soon as anyone perceived the odor he should raise his hand. I then unpacked the bottle in the front of the hall, poured the water over the cotton, holding my head away during the operation and started a stop-watch. While awaiting results I explained that I was quite sure that no one in the audience had ever smelled the chemical compound which I had poured out, and expressed the hope that, while they might find the odor strong and peculiar, it would not be too disagreeable to anyone. In fifteen seconds most of those in the front row had raised their hands, and in forty seconds the "odor" had spread to the back of the hall, keeping a pretty regular "wave front" as it passed on. About three-fourths of the audience claimed to perceive the smell, the obstinate minority including more men than the average of the whole. More would probably have succumbed to the suggestion, but at the end of a minute I was obliged to stop the experiment, for some on the front seats were being unpleasantly affected and were about to leave the room. No one in the audience seemed offended when it was explained that the real object of the experiment was the production of a hallucination.

Substitute "advertising" for "distilled water" and you have some idea of the powerful influence external factors—such as the incredible marketing muscle of the perfume industry and the scent industry in general—have over our noses. It's no less a force than that which persuaded a naked emperor that he was wearing beautiful clothes. If we accept the received aesthetics of the mass-produced perfume industry, we will live forever in a universe of "fruity florals," "floral florals," and other "florals" not found in nature that nevertheless last forever, immutable, on the skin.

On the other hand, our suggestibility means that exposure to authentic fragrances can have an almost instantly transformative effect on sensibility, conveying a richer, almost 3-D experience. When I passed around an unlabeled synthetic rose scent at a talk I gave on smell and the brain at the American Museum of Natural History, audience members guessed its identity as simply "lemon," "rose," or "soap." Smelling a real, natural rose scent elicited a wealth of responses: "rose," "lavender," "lemon," "balsam," "honey," "sandalwood," "orange," "patchouli," "fir," "sage," "mushrooms"—and that was only the beginning! The natural aroma showed its true complexity of character in the responses it evoked, touching a much wider swath of the imagination.

Developing an olfactory aesthetic—or, more deeply, a coherent and original personal aesthetic—is a long-term endeavor, for some the work of a lifetime, and among the more rewarding. It isn't a frivolous or self-indulgent exercise but a deeply serious one, as serious in its way as any we undertake. In a 1995 interview with the actress Anjelica Huston, music icon Leonard Cohen reflected on the "shabbiness" of our culture, the price of its frivolity and greed, and what might be the remedy for that:

> I think there is an appetite for seriousness. Seriousness is voluptuous, and very few people have allowed themselves the luxury of it. Seriousness is not Calvinistic, it's not a renunciation, it's the very opposite of that. Seriousness is the deepest pleasure that we can have.

On that note, I commend to you the serious, delirious journey into scent and the thrills of adventure, recognition, transcendence, strangeness, and beauty that await you there. I hope that this book, and the exceptional materials it has been my privilege to introduce to you, helps guide you on your way.

THE SHAPE OF A PERFUME

Creating a perfume is like creating anything. In the beginning every design possibility is open; any two or three essences might coexist. But once you have the central chord, a lot of doors and windows are shut. In perfume as in life, all is no longer possible.

There are so many aspects to consider as a perfume takes shape: What is the character of the blend? What is the texture? Is it sweet, bitter, heavy, light, pointed, dull? Does it unfold in layers? Creating a perfume that captivates often means marrying opposites, mimicking the oppositions that nature itself creates, as in jasmine's fecal-floral allure. Within a perfume composition, I am always balancing between the essences that play well with everyone versus those that have difficult but extraordinary personalities.

The essences being blended don't remain two distinct aromas in a blend. Facets of one aroma bond with facets of other aromas to create aromatic qualities none of them had to start with. The result often defies simple calculation, somehow resulting in more than the sum of the parts. As I have mentioned, I call this phenomenon "locking." In a good lock, a less expensive material might lend support to a costlier one or to a more fragile or elusive one. Or the flat vanilla aroma of benzoin might heighten and prolong the nascent sweetness of musky tobacco. In a bad lock, two essences that individually are not exceptionally sweet might result in an unpleasantly sugary effect. Understanding locking is incredibly useful in editing what has gone wrong in creating a perfume. The more you know about the different aromatic facets of a particular essence, the better you will be able to edit a blend, discovering where some aspect of one essence has locked with an aromatic aspect of another to create the undesired quality.

The ultimate expression of locking is the creation of an original fragrance, with its own identity, shape, and texture. When I am creating a perfume, dropping the essences one by one into the beaker and smelling after each addition, there comes a special moment when I recognize the birth of a separate and complete entity, when it becomes Honey Blossom or Sepia. This moment never ceases to be a thrill to me.

Perfumes have both shape and texture. *Shape* arises out of the direction in which the fragrance moves as the perfume evolves, like the shape of a melody. At some points—often the beginning—it may be sharp and pointy, at others it may be smooth. Often, as in a well-told story, the center is the highlight, rich with detail, and the denouement recedes. Sometimes the aromas lock together into a single, rounded shape, indivisible.

The *texture* of a perfume is its hallmark tone, the trait that best defines its character throughout its evolution. Is it sheer, rough, translucent, heavy? Smooth or rich, bright or dull, complex or sharp? The traits, and the vocabulary, overlap with those we associate with food and wine. Each material in the perfume makes its own textural contribution, and the overall texture can be shifted by pulling or prodding the various elements into the foreground or background—or sometimes pushing an ingredient offstage altogether.

Here is another arcane recipe, for "Elixir Monpou," from the *Techno-Chemical Receipt Book*, 1896:

Dissolve 120 drops of oil of peppermint, 40 each of oil of balm, oil of orange peel, rose essence, and orange blossom essence; 32 each of oil of mace and oil of cloves, and 60 of vanilla tincture, in 1½ gallons of rectified spirit of 90 per cent. Tr.; sweeten the solution with a syrup made of 7 pounds of sugar and 1¼ gallons of water. Color it rose-red.

FLORAL VARIATIONS

Jasmine has the uncanny ability to bring out the best in practically any other essence. There are two main varieties of jasmine absolute: sweeter and more floral jasmine grandiflorum, a vine that grows like a weed all over Northern California, and jasmine sambac. Both are rich and warm, heavy and fruity, intensely floral and slightly putrid (thanks to the indole); jasmine sambac is spicier and more "masculine." Good jasmine exhibits warmth and depth all the way through the drydown.

Rose absolute, congenial and unthreatening, can blend with any other essence without dominating. My favorite variety is a full-bodied and voluptuous rose from Turkey that has a staggering beauty and complexity. Rose absolute has the ability to smooth the rough edges in a blend and make a perfume more harmonious. Ylang-ylang is a creamy-sweet floral somewhat similar to jasmine, with banana-like notes. The finest distillation, called "extra," exhibits lift—the ability to add floral top notes as well as middle notes. Ylang-ylang will eagerly lock with jasmine and extend its aroma, which is useful because ylang-ylang is much less expensive. Tuberose, heavy and narcotic, has an earthy dirtiness in the drydown and also an almost phantom aroma of mint. In especially good versions of tuberose, there is an apricot facet as well. Orange flower absolute is cool, elegant, and intense—suave—and the best versions manage a seamless blend that merges citrus with floral, heavy with delicate, rich with fresh. The least overtly "floral" floral is geranium, which is rosy, light, and green and can be used to lighten heavy floral blends.

FLORALS FOR FRAGRANCE: All of these florals are middle notes of medium odor intensity; that said, there are nuances of difference—for example, jasmine has higher odor intensity than rose. All are absolutes, except ylang-ylang and geranium, which are essential oils.

The most expensive—and they are really pricey—are tuberose and orange flower. Florals are the backbone of perfume. Although conventional wisdom would indicate that florals are for women, I have found that men also love floral aromas for themselves. Natural florals, especially the indolic ones, have a handsome beauty with their layering of earthiness along with flowers, a tendency that can be underscored by leaning into woody, earthy, or spicy notes in blending.

ASMINE PERFUME

It has been said that two jasmines are better than one, and in this case using both sweet jasmine grandiflorum and spicy jasmine sambac creates a whole that equals much more than the sum of its parts. I think of it as having jasmine in the round, because of the many unfathomable trace notes that make up each jasmine absolute.

See pages 69–70 and 72–73 for basic blending procedures.

SOLID PERFUME

    8 milliliters jojoba oil

    Heaping ½ teaspoon grated beeswax

    7 drops jasmine grandiflorum absolute

    5 drops jasmine sambac absolute

    10 drops blood orange

    0.1 gram (small pea–size piece) lavender concrete

BODY OIL

20 milliliters jojoba oil

6 drops jasmine grandiflorum absolute

4 drops jasmine sambac absolute

2 drops geranium essential oil

5 drops blood orange essential oil

OIL-BASED PERFUME

10 milliliters fractionated coconut oil

6 drops jasmine grandiflorum absolute

6 drops jasmine sambac absolute

2 drops geranium essential oil

5 drops blood orange essential oil

ALCOHOL-BASED PERFUME

8 milliliters perfume alcohol

BASE NOTES

5 drops ambergris

4 drops labdanum absolute

0.1 gram (small pea–size piece) lavender concrete

MIDDLE NOTES

2 drops geranium essential oil

5 drops jasmine grandiflorum absolute

3 drops jasmine sambac absolute

TOP NOTES

3 drops geraniol natural isolate

3 drops blood orange essential oil

3 drops grand fir essential oil

With the two jasmines as the centerpiece, every other element is in service to them. The ambery labdanum and soapy lavender concrete provide a warm, rounded, shimmering foundation, thanks to the presence of the ambergris. Geranium in the middle notes provides a fresh counterpoint to the heady jasmines and connects to the greenness of the lavender concrete. The light rose aroma of geraniol with the jammy brightness of the fir in the top note mixes beautifully with the voluptuous raspberry splashiness of the blood orange.

FLORALS FOR FLAVOR: Rose brings a luscious creaminess to desserts, but when it is paired with ginger—leaning more to ginger than to rose—the flavors marry into something bright, floral, and spicy, with a hint of lemon. Rose and jasmine combine wonderfully to flavor teas, ice cream, pudding, shortbread (adapt the recipe in Chapter Four), and chocolate, either as a combination unto themselves or with spices and citruses. Jasmine is sultry by itself in pudding or tea, but paired with a hint of the coolness of mint it is clean and unexpected.

DEANA SIDNEY'S JASMINE-AMBERGRIS CHOCOLATE

Another wonderfully simple and revelatory recipe from Deana Sidney's exceptionally interesting blog, *Lost Past Remembered*.

6 ounces boiling water

1 ounce (about ¼ cup) grated 100% cacao chocolate

Small pea–size piece of ambergris

1½ teaspoons sugar or honey

¼ teaspoon vanilla extract

1 scant drop jasmine absolute (grandiflorum or sambac)

To the boiling water add the grated chocolate and stir until smooth. Mash the ambergris into the sugar or honey and add it to the chocolate. Stir to blend. If you have a cappuccino maker, give the mixture it a minute with the steamer; or heat it in a double boiler, whisking until it is foamy. You may also cool the mixture and store it, covered, on the counter overnight. By morning it will have developed a velvety texture that you may want to drink as is; or reheat it once more with a cappuccino maker or in a double boiler and whisk as instructed above. Add the vanilla extract and jasmine, whisk to blend, then pour into two small espresso cups (or one large cup just for you!), taking care not to leave behind any waxy specks of ambergris.

ACKNOWLEDGMENTS

I am particularly indebted to Harold McGee for his invaluable comments on my manuscript and for his precious friendship. I am also indebted to Daniel Patterson for his dear friendship and for his suggestion that I include "something green." Thanks to Ross Urerre for sharing his vast knowledge of incense. I am deeply grateful to William T. Vollmann, Chris Chapman, Heather Ive, and Andy Eales for their long friendship and their great kindnesses to me. Thanks also to my beloved stepchildren, Devon Curry-Leech, master pourer, who scanned my research and all the art that appears in the book, and Lauralyn Curry-Leech, whose hard work and keen design sense are such a help at Aftelier. To my daughter, Chloe Aftel, who is so far away: I will always treasure the years we had together.

Thanks to the amazingly creative, dedicated, and meticulous people at Riverhead: publisher Geoff Kloske for his support and attention; design team Helen Yentus, Claire Vaccaro, Janet Hansen, and Nicole LaRoche for the gorgeous package, inside and out; Maureen Sugden for deft copyediting and tweaking; Jynne Martin, Kate Stark, and their respective publicity and marketing crews for working tirelessly and imaginatively to bring before the world a book that is incredibly dear to my heart; and Wendy Pearl and the rest of the Penguin sales team for spreading the gospel.

Above all, I want to thank my best friend and editor, Becky Saletan, for making every sentence better and every aspect of creating this book a joy. And finally, to my husband, Foster Curry, who helped with every sentence in this book, thank you for being the best thing that ever happened to me and for bringing your formidable intelligence, grace, good humor, and generosity to every moment of our life together.

NOTES

CHAPTER ONE. A NEW NOSE

3   *"Odors have a power of persuasion":* Patrick Süskind, *Perfume* (New York: Penguin Books, 1986), 86.

4   *"And so he would now study perfumes":* Oscar Wilde, *The Picture of Dorian Gray* (New York: Penguin Classics, 2009), 129.

9   *He painted noses that "possess a will of their own":* Michael Taylor, *Rembrandt's Nose* (New York: Distributed Art Publishers, 2007), 20–21.

11   *"In our early, fishier version":* Diane Ackerman, *A Natural History of the Senses* (New York: Vintage, 1990), 20.

12   *"Smell seems to be the sense of singularity":* Michel Serres, *The Five Senses* (New York: Continuum, 2008), 169–70.

13   *Before the discovery of germs:* William Ian Miller, *The Anatomy of Disgust* (Cambridge, MA: Harvard University Press, 1997), 66.

13   *"Perfume would seem to be one of the elements":* Richard Le Gallienne, *The Romance of Perfume* (New York: Richard Hudnut, 1928), 6.

14   *"Spiced wine mingled with frankincense":* Susan Ashbrook Harvey, *Scenting Salvation* (Berkeley: University of California Press, 2006), 32.

14   *"It is precisely because of this inevitable familiarity":* Paul Freedman, *Out of the East: Spices and the Medieval Imagination* (New Haven: Yale University Press, 2008), 81.

15   *"I remember with gratitude the moment":* Serres, *Five Senses,* 155.

16   *"When an Ongee wishes to refer":* Constance Classen, *Worlds of Sense* (London: Routledge, 1993), 126–28.

21   *"Even though you're smelling a series":* Harold McGee, e-mail message to author, Dec. 16, 2013.

21   *"It's all good fun and marketing":* Avery Gilbert, *What the Nose Knows* (New York: Crown, 2008), 108.

22 *"a more convenient version of sticking your head out the window":* Harold McGee, e-mail message to author, Dec. 16, 2013.

23 *For safety guidelines with essential oils:* Robert Tisserand, *Essential Oil Safety: A Guide for Health Care Professionals,* 2nd ed. (Edinburgh: Churchill Livingstone, 2013).

25 *"The secret of cooking":* Patience Gray, *Honey from a Weed* (New York: Harper & Row, 1987), 94.

25 *"by reason that the flavour and scent":* William Terrington, *Cooling Cups and Dainty Drinks* (London: George Routledge, 1869), 209.

CHAPTER TWO. A TASTE FOR ADVENTURE

33 *"It is impossible to overstate":* Paul Freedman, *Out of the East: Spices and the Medieval Imagination* (New Haven: Yale University Press, 2008), 225.

33 *Spices and perfumes are mentioned in the records of ancient Sumer:* Shimshon Ben-Yehoshua, Carole Borowitz, and Lumír Ondřej Hanuš, "Frankincense, Myrrh, and Balm of Gilead: Ancient Spices of Southern Arabia and Judea," *Horticultural Reviews,* vol. 39, ed. Jules Janick (Hoboken, NJ: Wiley-Blackwell, 2012), 4.

33 *"Indeed, simple, unmixed scents appear to":* Edward H. Schafer, *The Golden Peaches of Samarkand: A Study of T'ang Exotics* (Berkeley: University of California Press, 1963), 159.

35 *"And yet humans have come to prize these weapons":* Harold McGee, *On Food and Cooking,* rev. ed. (New York: Scribner, 2004), 389.

35 *"The Arabians say that the dry sticks":* Herodotus, quoted in Andrew Dalby, *Dangerous Tastes: The Story of Spices* (Berkeley: University of California Press, 2000), 37.

36 *The Egyptians used cinnamon in mummification:* Jack Turner, *Spice: The History of a Temptation* (New York: Knopf, 2004), 230.

37 *Native bark collectors, understanding that their livelihood was at stake:* W. M. Gibbs, *Spices and How to Know Them* (Buffalo, NY: Matthews-Northrup Works, 1909), 82–85.

37 *The birds, however, were beyond Dutch control:* Ibid., 10.

38 *"While the ordinary China cassia":* Ibid., 88.

38 *"The quality of the bark:"* Ibid., p. 91.

38 *Moreover, branches exposed to direct sunlight:* Ibid.

39 *In the first stage of the harvest:* Lauren Rayner, "The Cinnamon Peeler's Life," Jan. 28, 2013, *The Cultureist,* www.thecultureist.com/2013/01/28/the-cinnamon-peeler-life-sri-lanka-madu-ganga.

42 *Hippocrates prescribed pepper:* J. Innes Miller, *The Spice Trade of the Roman Empire: 29 B.C. to A.D. 641* (Oxford: Clarendon Press, 1969), 2.

43 *"Spices must be mixed":* Ibid., 3–4.

43   *An antidote prepared in 80 B.C. for King Mithridates VI ...* "When these were ground": Ibid., 5.

44   *Regulations for perfumers in Constantinople:* From *The Book of the Eparch*, in Andrew Dalby, *Flavours of Byzantium* (Totnes, UK: Prospect Books, 2003), 40, quoted by Paul Freedman, *Out of the East*, 121.

45   *"The medieval spice merchant or apothecary":* Freedman, *Out of the East*, 119.

45   *"Postmortem surveys of London grocers' shops":* Ibid., 120.

47   *"To the consumer of spices, this should be said":* Gibbs, *Spices and How to Know Them*, 12.

48   *As articles of luxury, comparable with gems and silk:* Miller, *Spice Trade of the Roman Empire*, vii.

48   *"Not all consumers would have worked out in detail":* Freedman, *Out of the East*, 225.

49   *He likewise reported that frankincense:* Ibid., 135.

50   *"Retailers often emphasize or exaggerate the difficulty":* Ibid.

50   *"a perfume of green tamarind trees":* Charles Baudelaire, "Exotic Perfume," *The Flowers of Evil*, trans. Keith Waldrop (Middletown, CT: Wesleyan, 2006), 34.

52   *"It is fantastic to work beside Yves":* Hamish Bowles, "*Vogue* Remembers Loulou de La Falaise," Nov. 7, 2011, www.vogue.com/vogue-daily/article/vogue-remembers -loulou-de-la-falaise.

53   *"The conquest of the superfluous":* Gaston Bacheland, *The Psychoanalysis of Fire* (Boston: Beacon Press, 1968), p. 38.

53   *"No vital industry depended on spices":* John Keay, *The Spice Route: A History* (Berkeley: University of California Press, 2006), xii–xiii.

54   *"The opposite of luxury is not poverty":* Gabrielle "Coco" Chanel to photographer Cecil Beaton in 1966, http://davynedial.blogspot.com/2011/10/on-luxury-vs -vulgarity.html.

54   *"Yesterday's banquet ingredient":* Timothy Morton, *The Poetics of Spice: Romantic Consumerism and the Exotic* (Cambridge, UK: Cambridge University Press, 2000), 25.

CHAPTER THREE. THERE'S NO SMELL LIKE HOME

81   *"When we go to live in the house of memory":* Gaston Bachelard, "The Oneiric House," in *On Poetic Imagination and Reverie* (Putnam, CT: Spring Publications, 2005), 98.

82   *Fabulous assemblages of formulas, advice, and information:* William Eamon, *Science and the Secrets of Nature: Books of Secrets in Medieval and Early Modern Culture* (Princeton, NJ: Princeton University Press, 1994), 16.

82   *The most famous Book of Secrets:* Ibid., 359.

83   *"Pharmacists had to know":* Paul Freedman, *Out of the East: Spices and the Medieval Imagination* (New Haven: Yale University Press, 2008), 67.

83 *"We do not take very seriously the claim"*: Eamon, *Science and the Secrets of Nature*, 4–5.

85 *"Recipes collapse lived experience"*: Ibid., 360.

86 *"The Greeks called this type of knowledge metis"*: Ibid., 281.

87 *"A distinction has often been drawn"*: Pamela H. Smith, *The Body of the Artisan: Art and Experience in the Scientific Revolution* (Chicago: University of Chicago Press, 2004), 6–7.

87 *"How can a carpenter have any other book"*: Ibid., 85.

88 *Paracelsus ranked "doing" above "knowing"*: Ibid., 87.

90 **"Dioscorides** *saith it hath an heating"*: James E. Landing, *American Essence: A History of the Peppermint and Spearmint Industry in the United States* (Kalamazoo, MI: Kalamazoo Public Museum, 1969), 4–5.

91 *The Shoshone and the Paiute treated gas with mint tea:* University of Michigan, Native American Ethnobotany online database, http://herb.umd.umich.edu.

92 *In the Mediterranean the reference is to the traditional spearmint:* Ernest Small, *North American Cornucopia: Top 100 Indigenous Food Plants* (Boca Raton, FL: CRC, 2014), 705.

92 *In much of Central America:* Wikipedia main article on yerba buena, http://en.wikipedia.org/wiki/Yerba_buena.

92 *"Usually a free-lance"*: Richardson Wright, *Hawkers and Walkers in Early America: Strolling Peddlers, Preachers, Lawyers, Doctors, Players, and Others from the Beginning to the Civil War* (Philadelphia: Lippincott, 1927), 56–57.

94 *Born in 1817, he started out as a peddler of foodstuffs:* Eamon, *Science and the Secrets of Nature*, 359.

94 *Eventually Chase settled in Ann Arbor, Michigan:* Ibid.

95 *"Merchants and Grocers who retail vinegar"*: A. W. Chase, *Dr. Chase's Recipes; or, Information for Everybody*, rev. ed. (Chicago: Thompson & Thomas, 1903), 37.

96 *"soot scraped from a chimney"*: Ibid., 79.

97 *One pioneer in this industry:* Landing, *American Essence*, 45.

98 *an ambitious and innovative peppermint grower, Albert M. Todd:* Ibid., 67.

98 *In 1907 a report in the* **Farm Journal** *featured Todd's operation:* Walter E. Andrews, "Big Peppermint Farm—A Peculiar Industry Conducted by a Remarkable Man," *Farm Journal* (Philadelphia: Wilmer Atkinson) 31, no. 4 (Apr. 1907): 212.

99 *William Wrigley Jr., who had arrived in Chicago in 1891:* Landing, *American Essence*, 77.

100 *Weeds growing beside mint in the fields:* Ibid., 185.

100 *Their products were initially based:* Gary Reineccius, ed., *Source Book of Flavors*, 2nd ed. (New Delhi: CBS, 1997), 2.

101 *Eventually the secrets of their formulations became widely known:* Ibid.

101 *It detailed some of the herbs and spices:* Joseph Merory, *Food Flavorings: Composition, Manufacture, and Use* (Westport, CT: AVI, 1960), 201–336 passim.

102 *Food Additives Amendment of 1958:* U.S. Food and Drug Administration, Generally Recognized As Safe (GRAS), www.fda.gov/Food/IngredientsPackaging Labeling/GRAS.

104 *"A bowl of roses in a drawing-room":* Daphne du Maurier, *Rebecca* (New York: Harper, 2006), 30–31.

105 *"For our house is our corner of the world":* Gaston Bachelard, *The Poetics of Space* (Boston: Beacon Press, 1994), 4.

106 *"delightful homes set in ample gardens":* Charles Keeler, *Berkeley—Yesterday, Today, and Tomorrow,* pamphlet reprinted from article in *Berkeley Daily Gazette,* June 18, 1927.

106 *"the deepest and most permanent effect":* Leonard Woolf, *Downhill All the Way: An Autobiography of the Years 1919 to 1939* (San Diego: Harcourt Brace Jovanovich, 1975), 14.

107 *"If I were asked to name":* Bachelard, *Poetics of Space,* 6.

108 *"I never saw this strange dwelling again":* Rainer Maria Rilke in Bachelard, *Poetics of Space,* 57.

108 *"It is as though something fluid had collected ":* Bachelard, *Poetics of Space,* 57.

109 *"For the real houses of memory":* Ibid., 13.

109 *"We are always assailing ourselves":* Richard Todd, *The Thing Itself: On the Search for Authenticity* (New York: Riverhead, 2008), 34.

110 *"Diderot's regrets were prompted":* Juliet B. Schor, *The Overspent American: Why We Want What We Don't Need* (New York: HarperPerennial, 1999), 145.

115 *"It is said that this medicated":* Arnold James Cooley, *The Toilet and Cosmetic Arts in Ancient and Modern Times* (London: Robert Hardwicke, 1866), 750–51.

CHAPTER FOUR. REACHING FOR TRANSCENDENCE

125 *"Trees connect the three layers of the cosmos":* Jean Chevalier and Alain Gheerbrant, *A Dictionary of Symbols* (Cambridge, MA: Basil Blackwell, 1994), 1026–27.

126 *"Trees are sanctuaries":* Hermann Hesse, *Wandering: Notes and Sketches,* trans. James Wright (New York: Farrar, Straus & Giroux, 1972).

126 *In addition to turning their home into a golden temple:* Ovid, *Metamorphoses,* trans. A. D. Melville (Oxford: Oxford University Press, 1998), 8.616–724.

127 *"I am certain all sacred buildings":* John Fowles, *The Tree* (New York: Harper-Collins, 2010), 58.

128 *"It is not for nothing that the ancestors":* Ibid., 60.

129 *"These were burned to produce a pleasant fragrance":* Silvio A. Bedini, *The Trail of Time: Time Measurement with Incense in East Asia* (New York: Cambridge University Press, 1994), 26.

130 *"The acknowledged superiority of the Indochinese aromatics":* Edward H. Schafer, *The Golden Peaches of Samarkand: A Study of T'ang Exotics* (Berkeley: University of California Press, 1963), 158–59.

130   *There was a special incense for communicating with the dead:* Bedini, *Trail of Time*, 46.

130   *"Buddhist books were permeated with aromatic images":* Schafer, *Golden Peaches of Samarkand*, 157.

131   *"According to Japanese legend":* Bedini, *Trail of Time*, 46.

131   *In 78 B.C., the funeral of the dictator Sulla:* Caroline Singer, "The Incense Kingdoms of Yemen," in David Peacock and David Williams, eds., *Food for the Gods: New Light on the Ancient Incense Trade* (Oxford: Oxbow Books, 2007), 21.

132   *"A godly steam, and fit for godly nostrils":* Lucian, *The Works of Lucian of Samosata*, trans. H. W. Fowler and F. G. Fowler (Oxford: Clarendon Press, 1905), 125.

132   *"It might be as simple as frankincense alone":* Susan Ashbrook Harvey, *Scenting Salvation: Ancient Christianity and the Olfactory Imagination* (Berkeley: University of California Press, 2006), 12.

132–33   *"It does not live on fruit or flowers":* Ovid, quoted in Shimshon Ben-Yehoshua, Carole Borowitz, and Lumír Ondřej Hanuš, "Frankincense, Myrrh, and Balm of Gilead: Ancient Spices of Southern Arabia and Judea," *Horticultural Reviews*, vol. 39, ed. Jules Janick (Hoboken, NJ: Wiley-Blackwell, 2012), 29.

133   *the rabbis of the late Roman and early Byzantine periods:* Deborah Green, *The Aroma of Righteousness* (University Park: Pennsylvania State University Press, 2011), 4.

133   *"And when the sons of this man come":* Ibid., 181–91.

134   *In Exodus, God orders Moses:* Paul Freedman, *Out of the East: Spices and the Medieval Imagination* (New Haven: Yale University Press, 2008), 78.

135   *"Christian interpretations considered the constant evocation of perfumes":* Ibid., 77.

135   *"The men in charge of the fire started to light it":* Harvey, *Scenting Salvation*, 12.

136   *"And by his smelling in awe of the Lord":* Green, *Aroma of Righteousness*, 112–13.

136   *"In the overall context of the passage":* Ibid., 113.

137   *As many as twenty-five species of the **Boswellia** tree:* Ben-Yehoshua, Borowitz, and Hanuš, "Frankincense, Myrrh, and Balm of Gilead," 30.

137   **Boswellia sacra,** *historically the most prized of the frankincense trees:* Ibid., 10.

138   *To harvest the frankincense, the bark was shaved:* Schafer, *Golden Peaches of Samarkand*, 170.

139   *It is burned for forty days following the birth of a child:* Ben-Yehoshua, Borowitz, and Hanuš, "Frankincense, Myrrh, and Balm of Gilead," 12.

139   *Researchers at Johns Hopkins University:* Federation of American Societies for Experimental Biology, "Burning Incense Is Psychoactive: New Class of Antidepressants Might Be Right Under Our Noses," ScienceDaily, June 2008, www.sciencedaily.com/releases/2008/05/080520110415.htm.

139   *"separates the 'brain' of the cancerous cell":* Jeremy Howell, "Frankincense: Could It Be a Cure for Cancer?" Middle East Business Report, BBC World News, February 9, 2010, http://news.bbc.co.uk/2/hi/middle_east/8505251.stm.

139   *the longtime use of frankincense in treating arthritis:* Cardiff University, "A Wise Man's Treatment for Arthritis—Frankincense," Cardiff University News Centre,

June 1, 2011, www.cardiff.ac.uk/news/mediacentre/mediareleases/y2011/6844
.html.

140 *"The Egyptian word for myrrh"*: Ben-Yehoshua, Borowitz, and Hanuš, "Frankincense, Myrrh, and Balm of Gilead," 6.

140 *the most expensive perfumery material in the world:* Trygve Harris, "Agarwood—
Gem of Truth," Enfleurage.com, www.enfleurage.com/pages/Agarwood%
252d%252dGem-of-Truth.html.

140 *In the wild forest, the infection occurs in fewer than 10 percent:* Schafer, *Golden Peaches
of Samarkand,* 163.

141 *It was also tinctured in wine and used in Chinese medicine:* Ibid., 164.

141 *It helped that Malay:* Ibid., 34–35.

142 *In Japanese the word for "to sniff":* Bedini, *Trail of Time,* 41.

143 *In the Far East, for example, incense was used to mark time:* Ibid., 157.

143 *"Clocks in public buildings, particularly railroad stations":* Ibid., 6.

144 *"One day as he was passing through a rural region":* Ibid., 18.

145 *the period of engagement of a geisha:* Ibid., 180–84.

147 *"I would like to consider further those moments":* Agnes Martin, "On the Perfection
Underlying Life," in *Writings/Schriften* (Winterthur, Switzerland: Kunstmuseum
Winterthur; Stuttgart: Cantz, 1992), 68.

149 *"Reindeer moss, carefully picked over":* Arnold James Cooley, *The Toilet and Cosmetic
Arts in Ancient and Modern Times* (London: Robert Hardwicke, 1866), 622.

151 *"As frankincense burns in the censer":* Aida S. Kanafani, *Aesthetics and Ritual in the
United Arab Emirates: The Anthropology of Food and Personal Adornment Among
Arab Woman* (Syracuse, N.Y.: Syracuse University Press, 1984), 40.

CHAPTER FIVE. CURIOUS AND CURIOUSER

161 *"For some reason, it seems that whenever":* Steffen Arctander, *Perfume and Flavor
Materials of Natural Origin* (Elizabeth, NJ: Steffen Arctander, 1960), 422.

162 *"Striking visual demonstrations of the new valorization of curiosity":* William Eamon,
*Science and the Secrets of Nature: Books of Secrets in Medieval and Early Modern
Culture* (Princeton, NJ: Princeton University Press, 1994), 223–24.

163 *The items on display were the marvels of nature . . . "Natural Objects":* Ibid., 315.

163 *A modern version of the cabinet of curiosity is Pinterest:* Benjamin Breen, "Cabinets
of Curiosity: The Web as Wunderkammer," *The Appendix,* Nov. 28, 2012, https://
theappendix.net/blog/2012/11/cabinets-of-curiosity:-the-web-as-wunder
kammer.

165 *"Scent rolling is probably a way for wolves":* Sophia Yin, "Scent Rolling: Why Do
Dogs Like to Roll in Smelly Scents?" *Dr. Yin's Animal Behavior and Medicine Blog,*
June 30, 2011, http://drsophiayin.com/videos/tag/training%20tips/P50.

166 *Scent marking is used primarily for social communication:* D. Michael Stoddart,

Mammalian Odours and Pheromones, Studies in Biology, no. 73 (London: Edward Arnold, 1976), 46.

166 *"This begins, of course, as simple elimination":* Lyall Watson, *Jacobson's Organ and the Remarkable Nature of Smell* (New York: Norton, 2000), 40–41.

166 *The most prolific utilizer of scent:* Ibid., 41.

167 *When it is riled, it draws attention:* Stoddart, *Mammalian Odours and Pheromones*, 47.

167 *The secretion is stored in two sacs:* William F. Wood, "The History of Skunk Defense Secretion Research," *Chemical Educator* (New York: Springer-Verlag, 2000), 5, no. 3.

168 *"Chances were slim that I would get a chance":* George Clawson, *Trapping and Tracking* (New York: Winchester Press, 1977), 23–25.

169 *An intrepid researcher in the 1920s:* Austin H. Clark, "Fragrant Butterflies," *Smithsonian Institution Annual Report*, Washington, D.C., 1926.

170 *A fifteenth-century medical handbook:* Paul Freedman, *Out of the East: Spices and the Medieval Imagination* (New Haven: Yale University Press, 2008), 14–15.

172 *"Now this ambergris is a very curious substance":* Herman Melville, *Moby-Dick, or, The Whale* (New York: Penguin, 2009), chap. 92.

173 *In 2006, Loralee and Leon Wright:* "Couple Finds 32-Pound Hunk of Ambergris Worth over $300,000," The Exploding Whale, Jan. 25, 2006, http://theexploding whale.com/archives/2006/01/couple-finds-ambergris.

174 *Yet ambergris has been an article of trade:* Karl H. Dannenfeldt, "Ambergris: The Search for Its Origins," *Isis* 73, no. 3 (Sept. 1982): 382–97, www.jstor.org/stable/231442, p. 383.

174 *"The ambergris burned in the bellies":* Ibid.

175 *Ambergris was first introduced into medicine:* Ibid., 382.

176 *The German Georgius Everhardus Rumphius:* Georgius Everhardus Rumphius, *The Ambonese Curiosity Cabinet*, trans., ed., annot., with an intro. by E. M. Beekman (New Haven, CT: Yale University Press, 1999), 293.

179 *And the nineteenth-century author Pierre Lacour:* Pierre Lacour, *The Manufacture of Liquors, Wines, and Cordials, Without the Aid of Distillation* (New York: Dick & Fitzgerald, 1853), 20.

179 *Italian courtiers would sprinkle ground ambergris:* Andrew Dalby, *Dangerous Tastes: The Story of Spices* (Berkeley: University of California Press, 2000), 68, 146.

180 *famous gastronome Jean Anthelme Brillat-Savarin:* Jean Anthelme Brillat-Savarin, in Prosper Montagné, *Larousse Gastronomique: The Encyclopedia of Food, Wine and Cooking* (New York: Crown, 1961).

181 *The word "musk" is from the Sanskrit:* Volker Homes, *On the Scent: Conserving Musk Deer—The Uses of Musk and Europe's Role in Its Trade* (Brussels: Traffic Europe, 1999), 34.

182 *"The scent's mystical properties were so highly valued":* Stephen Fowler, "Musk: An

Essay," *Juice*, no. 3 (1995) www.pheromonetalk.com/pheromones-22158-post1 .html.

182 *Musk deer weigh twenty-five pounds:* Ibid.

182 *By the beginning of this century:* Ibid.

183 *efforts in China to raise the endangered Alpine musk deer:* Meng Xiuxiang et al., "Quantified Analyses of Musk Deer Farming in China: A Tool for Sustainable Musk Production and Ex Situ Conservation," *Asian-Australasian Journal of Animal Sciences* 24, no. 10 (Oct. 2011): 1473–82, http://koreascience.or.kr/article/ArticleFullRecord.jsp?cn=E1DMBP_2011_v24n10_1473.

184 *Its various species are native to Africa:* Takele taye Desta, "The African Civet Cat (*Viverra civetta*) and Its Life-Supporting Role in the Livelihood of Smallholder Farmers in Ethiopia," Tropentag 2009, University of Hamburg, Conference on International Research on Food Security, Natural Resource Management and Rural Development, Oct. 6–8, 2009.

184 *"Probably no foreign animal":* Karl H. Dannenfeldt, "Europe Discovers Civet Cats and Civet," *Journal of the History of Biology* 18, no. 3 (Autumn 1985): 430–31, www.jstor.org/stable/4330947.

184 *These days civet paste from farmed animals:* Ibid.

185 *Priests at a temple in India:* James McHugh, "The Disputed Civets and the Complexion of the God: Secretions and History in India," *Journal of the American Oriental Society* 132, no. 2 (Apr. 1, 2012): 245–73.

185 *Typically farmers raise them:* Desta, "African Civet Cat."

186 *In the nineteenth century, beavers were trapped:* Dietland Müller-Schwarze and Lixing Sun, *The Beaver: Natural History of a Wetlands Engineer* (Ithaca, NY: Comstock, 2003), 149.

187 *Rock hyraxes are native to Africa:* Andrew S. Carr, Arnoud Boom, and Brian M. Chase, "The Potential of Plant Biomarker Evidence Derived from Rock Hyrax Middens as an Indicator of Palaeoenvironmental Change," *Palaeogeography, Palaeoclimatology, Palaeoecology* 285 (2010): 321.

188 *found in abundance along the shores of the Red Sea:* James McHugh, "Blattes de Byzance in India: Mollusk Opercula and the History of Perfumery," *Journal of the Royal Asiatic Society* 23, no. 1 (Jan. 2013): 53–67.

189 *It is created by roasting crushed onycha shells:* Christopher McMahon, "Hina—India's Mystery Perfume," White Lotus Newsletter, 2000, www.whitelotus aromatics.com/newsletters/hina.

189 *"When broken into large chunks and laid on coals":* Rumphius, *Ambonese Curiosity Cabinet*, 125.

190 *"At Schiraz . . . we met a merchant":* Paolo Rovesti, *In Search of Perfumes Lost* (Venice: Blow Up/Marsilio Editori, 1980), 179–80.

191 *"Humans are enclosed in a powerful olfactory":* Stoddart, *Mammalian Odours and Pheromones*, 58.

NOTES

191 *"Lingering near the surface of skins":* Michel Serres, *The Five Senses* (New York: Continuum, 2008), 170–71.

CHAPTER SIX. SEDUCED BY BEAUTY

205 *"Beauty can be consoling":* Roger Scruton, *Beauty: A Very Short Introduction* (New York: Oxford University Press, 2011), xii.

206 *"Is Jasmine then the mystical Morn":* Donald McDonald, *Sweet-Scented Flowers and Fragrant Leaves: Interesting Associations Gathered from Many Sources, with Notes on Their History and Utility* (London: Sampson Low, Marston, 1895), 61.

206 *"the foul and the fragrant" are flip sides:* Alain Corbin, *The Foul and the Fragrant: Odor and the French Social Imagination* (Cambridge, MA: Harvard University Press, 1988).

208 *"My visit is shorter than a ghost's":* "Song of the Rose," *The Book of the Thousand Nights and One Night—Rendered from the Literal and Complete Version of Dr J. C. Mardrus; and Collated with Other Sources,* trans. E. Powys Mathers (London: Casanova Society, 1923).

211 *"Yet a small window exercised a discipline of proportion":* John O'Donohue, *Beauty: The Invisible Embrace* (New York: HarperPerennial, 2004), 36.

211 *"The main strategy of this intelligence":* Leonard Koren, *Wabi-Sabi for Artists, Designers, Poets & Philosophers* (Berkeley, CA: Stone Bridge Press, 1994), 72.

214 *"It is a wise and true language":* Michel Serres, *The Five Senses* (New York: Continuum, 2008), 171.

214 *"with the correct balance of allies":* James McHugh, *Sandalwood and Carrion: Smell in Indian Religion and Culture* (Oxford: Oxford University Press, 2012), 5.

215 *"The different marks I employ":* Henri Matisse, "Notes of a Painter" ("Notes d'un Peintre," from *La Grande Revue,* Paris, Dec. 25, 1908), in Jack Flam, ed., *Matisse on Art* (Berkeley: University of California Press, 1995).

215 *The medieval theologian and philosopher Alexander of Hales:* Umberto Eco, ed., *History of Beauty,* trans. Alastair McEwen (New York: Rizzoli, 2004), 148.

215 *"There's an aching poetry in things":* Robyn Griggs Lawrence, *The Wabi-Sabi House: The Japanese Art of Imperfect Beauty* (New York: Clarkson Potter, 2004), 21.

216 *"The physical decay or natural wear":* Andrew Juniper, *Wabi Sabi: The Japanese Art of Impermanence* (Tokyo: Tuttle, 2003), 106.

217 *"Everything in the universe":* Ibid, 27.

217 **Mono-no-aware** *is an ancient principle:* Boyé Lafayette De Mente, *Elements of Japanese Design* (Tokyo: Tuttle, 2006), 126.

217 *A formula for an incense known as "Bare Boughs":* Marcel Billot and F. V. Wells, *Perfumery Technology: Art: Science: Industry* (Chichester, UK: Ellis Horwood, 1975), 19.

218 *the Japanese appreciation for mutability:* Donald Keene, *The Pleasures of Japanese Literature* (New York: Columbia University Press, 1988), 20–21.

219 *"The soul is weighed in the balance":* O'Donohue, *Beauty: The Invisible Embrace*, 13.

220 *"Beauty touches and renews":* Ibid., 20.

221 *"a modern city without steel and glass":* Alessandra Codinha, "Talking Scents with Francis Kurkdjian," *Into the Gloss*, Jan. 3, 2013, http://intothegloss.com/2013/01/talking-scents-with-francis-kurkdjian.

221 *breaking down the profit-and-loss assumptions:* Barbara Thau, "Behind the Spritz: What Really Goes into a Bottle of $100 Perfume," DailyFinance, May 22, 2012, www.dailyfinance.com/2012/05/22/celebrity-perfume-cost-breakdown.

222 *"Songs, to me, were more important":* Bob Dylan, *Chronicles: Volume One* (New York: Simon & Schuster, 2005), 34.

222 *"The function of art work":* Agnes Martin, *Writings/Schriften* (Winterthur, Switzerland: Kunstmuseum Winterthur; Stuttgart: Cantz, 1992), 69.

223 *"When I think of art I think of beauty":* Agnes Martin in Arne Glimcher, *Agnes Martin: Paintings, Writings, Remembrances* (London: Phaidon Press, 2012), 168.

224 *"In the question of decoration":* Oscar Wilde, "The House Beautiful," in Tweed Conrad, ed., *Oscar Wilde in Quotation* (Jefferson, NC: McFarland, 2006), chap. 33, "Home Decor."

227 *"The following scheme for the production":* E. E. Slosson, "Shorter Communications and Discussions: A Lecture Experiment in Hallucinations," *Psychological Review* 6, no. 4 (July 1, 1899): 407–8.

229 *"I think there is an appetite for seriousness":* Leonard Cohen, interviewed by Anjelica Huston, *Interview*, Nov. 1995.

BIBLIOGRAPHY

BOOKS

Ackerman, Diane. *A Natural History of the Senses*. New York: Vintage, 1990.

Adams, William Howard. *On Luxury: A Cautionary Tale*. Washington, D.C.: Potomac Books, 2012.

Addis, Laird. *Of Mind and Music*. Ithaca, NY: Cornell University Press, 1999.

Aftel, Mandy. *Aroma: The Magic of Essential Oils in Food and Fragrance*. New York: Artisan, 2004.

————. *Essence and Alchemy: A Natural History of Perfume*. Layton, UT: Gibbs Smith, 2008.

————. *Scents and Sensibilities: Creating Solid Perfumes for Well-Being*. Layton, Utah: Gibbs Smith, 2005.

Akasoy, Anna, Charles Burnett, and Ronit Yoeli-Tlalim. *Islam and Tibet—Interactions Along the Musk Routes*. Burlington, VT: Ashgate, 2011.

Arctander, Steffen. *Perfume amd Flavor Materials of Natural Origin*. Elizabeth, NJ: Steffen Arctander, 1960.

Armstrong, John. *The Secret Power of Beauty: Why Happiness Is in the Eye of the Beholder*. London: Penguin Books, 2004.

Atchley, E. G. Cuthbert. *A History of the Use of Incense in Divine Worship*. New York: Longmans, Green, 1909.

Bachelard, Gaston. *Air and Dreams: An Essay on the Imagination of Movement*. Dallas: Dallas Institute, 1988.

————. *On Poetic Imagination and Reverie*. Putnam, CT: Spring Publications, 2005.

————. *The Poetics of Space*. Boston: Beacon Press, 1994.

————. *The Psychoanalysis of Fire*. Boston: Beacon Press, 1964.

Baudelaire, Charles. *The Flowers of Evil*. Trans. Keith Waldrop. Middletown, CT: Wesleyan, 2006.

Beasley, Henry. *The Druggist's General Receipt Book*. 6th ed. London: John Churchill and Sons, 1866.

Bedini, Silvio A. *The Trail of Time: Time Measurement with Incense in East Asia*. New York: Cambridge University Press, 1994.

Benjamin, Walter. *The Work of Art in the Age of Its Technological Reproducibility and Other Writings on Media*. Cambridge, MA: Harvard University Press, 2008.

Berry, Christopher. *The Idea of Luxury: A Conceptual and Historical Investigation*. New York: Cambridge University Press, 1994.

Billot, Marcel, and F. V. Wells. *Perfumery Technology: Art: Science: Industry*. Chichester, UK: Ellis Horwood, 1975.

Blackwell, Lewis. *The Life & Love of Trees*. San Francisco: Chronicle Books, 2010.

Boyd, Andrew. *Spirit of Air: A Complete Book of Incense*. InnerWitch Press, 2005.

Brannt, William, and William Wahl. *The Technochemical Receipt Book*. Philadelphia: Henry Carey Baird, 1896.

Buhner, Stephen Harold. *The Lost Language of Plants: The Ecological Importance of Plant Medicines to Life on Earth*. White River Junction, VT: Chelsea Green, 2002.

Burton, Robert. *The Language of Smell*. Boston: Routledge & Kegan Paul, 1976.

Carlisle, Janice. *Common Scents: Comparative Encounters in High-Victorian Fiction*. Oxford: Oxford University Press, 2004.

Chase, A. W. *Dr. Chase's Recipes; or Information, for Everybody*. Rev. ed. Chicago: Thompson & Thomas, 1903.

Chevalier, Jean, and Alain Gheerbrant. *A Dictionary of Symbols*. Cambridge, MA: Basil Blackwell, 1994.

Classen, Constance. *Worlds of Sense: Exploring the Senses in History and Across Cultures*. London: Routledge, 1993.

Clawson, George. *Trapping and Tracking*. New York: Winchester Press, 1977.

Conigliaro, Tony. *Drinks: Unravelling the Mysteries of Flavour and Aroma in Drink*. London: Ebury Press, 2012.

Cooley, Arnold J. *Handbook of Perfumes, Cosmetics, and Other Toilet Articles*. Philadelphia: Lippincott, 1873.

———. *The Toilet and Cosmetic Arts in Ancient and Modern Times*. London: Robert Hardwick, 1866.

Corbin, Alain. *The Foul and the Fragrant: Odor and the French Social Imagination*. Cambridge, MA: Harvard University Press, 1988.

Crawley, Ernest. *Oath, Curse, and Blessing*. London: Watts, 1934.

Dalby, Andrew. *Dangerous Tastes: The Story of Spices*. Berkeley: University of California Press, 2000.

Daston, Lorraine, and Katharine Park. *Wonders and the Order of Nature*. New York: Zone Books, 2001.

Davenne, Christine. *Cabinets of Wonder*. New York: Harry N. Abrams, 2012.

Day, Ivan. *Perfumery with Herbs*. London: Darton, Longman & Todd, 1979.

DeJean, Joan. *The Essence of Style*. New York: Free Press, 2005.

De Mente, Boyé Lafayette. *Elements of Japanese Design*. Tokyo: Tuttle, 2006.

Detienne, Marcel. *The Gardens of Adonis: Spices in Greek Mythology*. Princeton, NJ: Princeton University Press, 1972.

Dick, William B. *Encyclopedia of Practical Receipts and Processes*. New York: Dick & Fitzgerald, 1891.

Dorland, Wayne E., and James A. Rogers Jr. *The Fragrance and Flavor Industry*. Mendham, NJ: Wayne E. Dorland, 1977.

Dragoco Report, no. 5. Totowa, NJ: Dragoco, 1998.

du Maurier, Daphne. *Rebecca*. New York: Harper, 2006.

Dugan, Holly. *The Ephemeral History of Perfume: Scent and Sense in Early Modern England*. Baltimore: Johns Hopkins University Press, 2011.

Dylan, Bob. *Chronicles: Volume One*. New York: Simon & Schuster, 2005.

Eamon, William. *Science and the Secrets of Nature: Books of Secrets in Medieval and Early Modern Culture*. Princeton, NJ: Princeton University Press, 1994.

Eco, Umberto, ed. *History of Beauty*. Trans. Alastair McEwen. New York: Rizzoli, 2004.

Ellena, Jean-Claude. *The Diary of a Nose: A Year in the Life of a Parfumeur*. London: Penguin, 2012.

———. *Perfume: The Alchemy of Scent*. New York: Arcade, 2011.

Ellish, Aytoun. *The Essence of Beauty: A History of Perfume and Cosmetics*. London: Secker & Warburg, 1960.

Eslamieh, Jason. *Cultivation of Boswellia: Sacred Trees of Frankincense*. Phoenix: Book's Mind, 2011.

Ferguson, John. *Bibliographical Notes on Histories of Inventions and Books of Secrets*. London: Holland Press, 1981. Repr. of 1959 ed.

Ferry-Swainson, Kate. *Mint*. Boston: Tuttle, 2000.

Findlen, Paula. *Possessing Nature: Museums, Collecting, and Scientific Culture in Early Modern Italy*. Berkeley: University of California Press, 1994.

Flam, Jack, ed. *Matisse on Art*. Berkeley: University of California Press, 1995.

Fowles, John. *The Tree*. New York: HarperCollins, 2010.

Freedman, Paul. *Out of the East: Spices and the Medieval Imagination*. New Haven: Yale University Press, 2008.

Fritzsche Brothers. *Fritzsche's Manual: For the Manufacture of Liquors etc*. New York: Fritzsche Brothers, 1896.

Gazan, M. H. *Flavours and Essences: A Handbook of Formulae*. London: Chapman & Hall, 1936.

Gibbs, W. M. *Spices and How to Know Them*. Buffalo: Matthews-Northrup Works, 1909.

Gilbert, Avery. *What the Nose Knows*. New York: Crown, 2008.

Gildmeister, E., and Fr. Hoffman. *The Volatile Oils*. Milwaukee: Pharmaceutical Review, 1900.

Glimcher, Arne. *Agnes Martin: Paintings, Writings, Remembrances*. London: Phaidon Press, 2012.

Gopnik, Adam. *The Table Comes First: Family, Fragrance, and the Meaning of Food*. New York: Knopf, 2011.

Gray, Patience. *Honey from a Weed*. New York: Harper & Row, 1987.

Green, Deborah. *The Aroma of Righteousness*. University Park: Pennsylvania State University Press, 2011.

Groom, Nigel. *Frankincense and Myrrh: A Study of the Arabian Incense Trade*. London: Longman Group, 1981.

Guillén, Diego García, Agustín Albarracín, Elvira Arquiola, Sergio Erill, Luis Montiel, José Luis Peset, and Pedro Laín Entralgo. *History of Medicament*. Barcelona: Ediciones Doyma, 1985.

Hale, Jonathan. *The Old Way of Seeing: How Architecture Lost Its Magic (and How to Get It Back)*. Boston: Houghton Mifflin, 1994.

Harper, H. W. *Harper's Universal Recipe Book*. Boston: Geo. B. Oakes, 1869.

Harvey, Susan Ashbrook. *Scenting Salvation: Ancient Christianity and the Olfactory Imagination*. Berkeley: University of California Press, 2006.

Hesse, Hermann. *Wandering: Notes and Sketches*. Trans. James Wright. New York: Farrar, Straus & Giroux, 1972.

Highet, Juliet. *Frankincense: Oman's Gift to the World*. New York: Prestel, 2006.

Hiss, A. Emil, and Albert E. Ebert. *The New Standard Formulary*. Chicago: G. P. Engelhard, 1910.

Hort, Sir Arthur. *Theophrastus: Enquiry into Plants*. 5th ed., book 2. Cambridge, MA: Harvard University Press, 1953.

Hume, Nancy, ed. *Japanese Aesthetics and Culture: A Reader*. Albany: State University of New York Press, 1995.

Jung, Dinah. *An Ethnography of Fragrance: The Perfumery Arts of Adan/Lahj*. Leiden, Netherlands: Brill, 2011.

Juniper, Andrew. *Wabi Sabi: The Japanese Art of Impermanence*. Tokyo: Tuttle, 2003.

Kanafani, Aida S. *Aesthetics and Ritual in the United Arab Emirates: The Anthropology of Food and Personal Adornment Among Arab Women*. Syracuse, NY: Syracuse University Press, 1984.

Keay, John. *The Spice Route: A History*. Berkeley: University of California Press, 2006.

Keeler, Charles. *The Simple Home*. Santa Barbara, CA: Peregrine Smith, 1979. Repr. of 1904 ed.

Keene, Donald. *The Pleasures of Japanese Literature*. New York: Columbia University Press, 1988.

Kesseler, Rob, and Madeline Harley. *Pollen: The Hidden Sexuality of Flowers*. Buffalo, NY: Firefly Books, 2009.

Koide, Yukiko, and Kyoichi Tsuzuki. *Boro: Rags and Tatters from the Far North of Japan*. Tokyo: Aspect, 2009.

Koren, Leonard. *Wabi-Sabi for Artists, Designers, Poets & Philosophers*. Berkeley, CA: Stone Bridge Press, 1994.

Lacour, Pierre. *The Manufacture of Liquors, Wines, and Cordials, Without the Aid of Distillation*. New York: Dick & Fitzgerald, 1853.

Landing, James E. *American Essence: A History of the Peppermint and Spearmint Industry in the United States*. Kalamazoo, MI: Kalamazoo Public Museum, 1969.

Langenheim, Jean H. *Plant Resins: Chemistry, Evolution, Ecology, Ethnobotany*. Portland, OR: Timber Press, 2003.

Lawrence, Robyn Griggs. *The Wabi-Sabi House: The Japanese Art of Imperfect Beauty*. New York: Clarkson Potter, 2004.

Lawton, Barbara Perry. *Mints: A Family of Herbs and Ornamentals*. Portland, OR: Timber Press, 2002.

Lawton, John. *Silk, Scents, & Spice*. Paris: United Nations Educational, Scientific and Cultural Organization, 2004.

Lazennec, I. *Manuel de Parfumerie*. Paris: J.-B. Baillière et Fils, 1922.

Le Gallienne, Richard. *The Romance of Perfume*. New York: Richard Hudnut, 1928.

Lucas, A., and J. R. Harris. *Ancient Egyptian Materials and Industries*. Mineola, NY: Dover, 1999.

Lucian. *The Works of Lucian of Samosata*. Trans. H. W. Fowler and F. G. Fowler. Oxford: Clarendon Press, 1905.

Marcus, Greil. *The Old, Weird America: The World of Bob Dylan's Basement Tapes*. Rev. ed. New York: Picador, 2011.

Martin, Agnes. *Writings/Schriften*. Winterthur, Switzerland: Kunstmuseum Winterthur; Stuttgart: Cantz, 1992.

Martin, Geoffrey. *Perfumes, Essential Oils & Fruit Essences*. London: Crosby Lockwood and Son, 1921.

Mathers, E. Powys, trans. *The Book of the Thousand Nights and One Night—Rendered from the Literal and Complete Version of Dr. J. C. Mardrus; and Collated with Other Sources*. London: Casanova Society, 1923.

Matthews, Leslie G. *The Antiques of Perfume*. London: G. Bell & Sons, 1973.

May, Robert. *The Accomplisht Cook; or, The Art and Mystery of Cookery*. 5th ed. London: Dodo Press, 1685.

McDonald, Donald. *Sweet-Scented Flowers and Fragrant Leaves: Interesting Associations Gathered from Many Sources, with Notes on Their History and Utility*. London: Sampson Low, Marston, 1895.

McGee, Harold. *On Food and Cooking*. Rev. ed. New York: Scribner, 2004.

McHugh, James. *Sandalwood and Carrion: Smell in Indian Religion and Culture.* Oxford: Oxford University Press, 2012.

Melville, Herman. *Moby-Dick, or, the Whale.* New York: Penguin, 2009.

Merory, Joseph. *Food Flavorings: Composition, Manufacture, and Use.* Westport, CT: AVI, 1960.

Miller, Amy B. *Shaker Medicinal Herbs: A Compendium of History, Lore, and Uses.* Pownal, VT: Storey, 1998.

Miller, J. Innes. *The Spice Trade of the Roman Empire: 29 B.C. to A.D. 641.* Oxford: Clarendon Press, 1969.

Miller, William Ian. *The Anatomy of Disgust.* Cambridge, MA: Harvard University Press, 1997.

Mohr, Francis, and Theophilus Redwood. *Practical Pharmacy: The Arrangements, Apparatus, and Manipulations of the Pharmaceutical Shop and Laboratory.* Philadelphia: Lea and Blanchard, 1849.

Montagné, Prosper. *Larousse Gastronomique: The Encyclopedia of Food, Wine and Cooking.* 1st English ed. New York: Crown, 1961.

Morita, Kiyoko. *The Book of Incense: Enjoying the Traditional Arts of Japanese Scents.* Tokyo: Kodansha, 1992.

Morton, Timothy. *The Poetics of Spice: Romantic Consumerism and the Exotic.* Cambridge, UK: Cambridge University Press, 2000.

Müller-Schwarze, Dietland, and Lixing Sun. *The Beaver: Natural History of a Wetlands Engineer.* Ithaca, NY: Comstock, 2003.

Norman, Jill. *Herbs & Spices: The Cook's Reference.* New York: Dorling Kindersley, 2002.

Nussinovitch, Amos. *Plant Gum Exudates of the World: Sources, Distribution, Properties, and Applications.* Boca Raton, FL: CRC Press, Taylor & Francis, 2010.

O'Donohue, John. *Beauty: The Invisible Embrace.* New York: HarperPerennial, 2004.

Ohloff, Gunther, et al. *Scent and Chemistry: The Molecular World of Odors.* Zurich: VHCA, 2012.

O'Neil, Darcy. *Fix the Pumps.* Self-published, 2009.

Ovid. *Metamorphoses.* Trans. A. D. Melville. Oxford: Oxford University Press, 1998.

Parrish, Edward. *A Treatise on Pharmacy.* Ed. Thomas S. Wiegand. 5th ed. Philadelphia: Henry Lea's Son, 1884.

Parry, Ernest J. *Pitman's Common Commodities and Industries: Gums & Resins: Their Occurrence, Properties and Uses.* London: Sir Isaac Pitman & Sons, 1918.

Parry, John. *Spices,* vol. 2: *Morphology, Histology, Chemistry.* New York: Chemical, 1969.
———. *The Story of Spices.* New York: Chemical, 1953.

Peacock, David, and David Williams, eds. *Food for the Gods: New Light on the Ancient Incense Trade.* Oxford: Oxbow Books, 2007.

Pennacchio, Marcello, Lara Jefferson, and Kayri Havens. *Uses & Abuses of Plant-*

Derived Smoke: Its Ethnobotany as Hallucinogen, Perfume, Incense & Medicine. New York: Oxford University Press, 2010.

Piesse, Septimus. *Chimie des Parfum et Fabrication des Essences.* Paris: J.-B. Baillière et Fils, 1897.

Pine, Joseph II, and James Gilmore. *The Experience Economy: Work Is Theatre & Every Business a Stage.* Boston: Harvard Business Review Press, 1999.

Plat, Hugh. *Delightes for Ladies.* London: Crosby Lockwood and Son, 1948.

Pollan, Michael. *The Botany of Desire: A Plant's-Eye View of the World.* New York: Random House, 2001.

Pollard, H. B. C. *The Mystery of Scent.* London: Eyre and Spottiswoode, 1937.

Rack, John. *The French Wine and Liquor Manufacturer.* 4th ed. New York: Dick & Fitzgerald, 1868.

Reineccius, Gary, ed. *Source Book of Flavors.* 2nd ed. New Delhi: CBS, 1997.

Richie, Donald. *A Tractate on Japanese Aesthetics.* Berkeley, CA: Stone Bridge Press, 2007.

Roundell, Julia Anne, Elizabeth Tollemache, and Harry Roberts. *The Still Room.* London: John Lane, 1903.

Rovesti, Paolo. *In Search of Perfumes Lost.* Venice: Blow Up/Marsilio Editori, 1980.

Rumphius, Georgius Everhardus. *The Ambonese Curiosity Cabinet.* Trans., ed., with an intro. by E. M. Beekman. New Haven: Yale University Press, 1999.

Schafer, Edward H. *The Golden Peaches of Samarkand: A Study of T'ang Exotics.* Berkeley: University of California Press, 1963.

Schivelbusch, Wolfgang. *Tastes of Paradise: A Social History of Spices, Stimulants, and Intoxicants.* New York: Random House, 1992.

Schor, Juliet B. *The Overspent American: Why We Want What We Don't Need.* New York: HarperPerennial, 1999.

Scruton, Roger. *Beauty: A Very Short Introduction.* New York: Oxford University Press, 2011.

Segnit, Niki. *The Flavor Thesaurus: Pairings, Recipes and Ideas for the Creative Cook.* New York: Bloomsbury USA, 2010.

Serres, Michel. *The Five Senses.* New York: Continuum, 2008.

Sheperd, Gordon. *Neurogastronomy: How the Brain Creates Flavor and Why It Matters.* New York: Columbia University Press, 2012.

Small, Ernest. *North American Cornucopia: Top 100 Indigenous Food Plants.* Boca Raton, FL: CRC Press, 2014.

Smith, Pamela H. *The Body of the Artisan: Art and Experience in the Scientific Revolution.* Chicago: University of Chicago Press, 2004.

Spon, E., and F. N. Spon. *Workshop Receipts for Manufacturing and Scientific Amateurs.* London: E. & F. N. Spon, 1917.

Stoddart, D. Michael. *Mammalian Odours and Pheromones,* Studies in Biology, no. 73. London: Edward Arnold, 1976.

Stuckey, Barb. *Taste What You're Missing: The Passionate Eater's Guide to Why Good Food Tastes Good*. New York: Free Press, 2012.

Süskind, Patrick. *Perfume*. New York: Penguin, 1986.

Taylor, Michael. *Rembrandt's Nose: Of Flesh & Spirit in the Master's Portraits*. New York: Distributed Art Publishers, 2007.

Teiji, Itoh, et al. *Wabi Sabi Suki: The Essence of Japanese Beauty*. Tokyo: Cosmo Public Relations, 1993.

Terrington, William. *Cooling Cups and Dainty Drinks*. London: George Routledge, 1869.

Thomas, Dana. *Deluxe: How Luxury Lost Its Luster*. New York: Penguin, 2007.

Tisserand, Robert. *Essential Oil Safety: A Guide for Health Care Professionals*. 2nd ed.. Edinburgh: Churchill Livingstone, 2013.

Todd, Richard. *The Thing Itself: On the Search for Authenticity*. New York: Riverhead, 2008.

Turner, Jack. *Spice: The History of a Temptation*. New York: Knopf, 2004.

Verdoorn, Frans. *Vegetable Gums and Resins*. Waltham, MA: Chronica Botanica, 1949.

Waddington, Shelley. *Perfuming with Natural Isolates: The Complete Reference Manual*. Shelley Waddington, 2011.

Waite, Arthur Edward. *The Hermetic and Alchemical Writings of Paracelsus*, vol. 1, *Hermetic Chemistry*. Berkeley, CA: Shambhala, 1976.

————. *The Hermetic and Alchemical Writings of Paracelsus*, vol. 2, *Hermetic Medicine and Hermetic Philosophy*. Berkeley, CA: Shambhala, 1976.

Walker, Harlan, ed. *Food and the Memory: Proceedings of the Oxford Symposium on Food and Cookery 2000*. Devon, UK: Prospect, 2001.

Walter, Erich. *Manual for the Essence Industry*. New York: John Wiley & Sons, 1916.

Watson, Lyall. *Jacobson's Organ and the Remarkable Nature of Smell*. New York: Norton, 2000.

Wilde, Oscar. *The Picture of Dorian Gray*. New York: Penguin, 2009.

Williams, David G. *Perfumes of Yesterday*. Port Washington, NY: Micelle Press, 2004.

Winter, Ruth. *Scent Talk Among Animals*. New York: Lippincott, 1977.

Woolf, Leonard. *Downhill All the Way: An Autobiography of the Years 1919 to 1939*. San Diego: Harcourt Brace Jovanovich, 1975.

Wright, John. *Flavor Creation*. Carol Stream, IL: Allured, 2005.

Wright, Richardson. *Hawkers and Walkers in Early America: Strolling Peddlers, Preachers, Lawyers, Doctors, Players, and Others from the Beginning to the Civil War*. Philadelphia: Lippincott, 1927.

Zappler, Lisbeth, and Jean Zallinger. *The Natural History of the Nose*. Garden City, NY: Doubleday, 1976.

ARTICLES AND REPORTS

Abrahams, Harold J. "Onycha, Ingredient of the Ancient Jewish Incense: An Attempt at Identification." *Economic Botany* 33, no. 2 (Apr.–June 1979): 233–36.

Akasoy, Anna. "Along the Musk Routes: Exchanges Between Tibet and the Islamist World." *Asian Medicine* 3, no. 2 (2007): 217–40.

Albert Vieille. "African Stone." Albert Vieille Report, www.albertvieille.com/en/upload/210313_101247_PEEL_QHnNDgvH.pdf.

Andrews, Walter E. "Big Peppermint Farm—A Peculiar Industry Conducted by a Remarkable Man." *Farm Journal* (Philadelphia: Wilmer Atkinson) 31, no. 4 (Apr. 1907).

Aryal, Achyut. "Himalayan Musk Deer in Annapurna Conservation Area, Nepal." *Tigerpaper* 33, no. 2 (Apr.–June 2006), www.forestrynepal.org/images/publications/musk.pdf.

BBC News. "Whale 'Vomit' Sparks Cash Bonanza." BBC News Asia-Pacific, Jan. 24, 2006, http://news.bbc.co.uk/2/hi/asia-pacific/4642722.stm.

Ben-Yehoshua, Shimson, Carole Borowitz, and Lumír Ondřej Hanuš. "Frankincense, Myrrh, and Balm of Gilead: Ancient Spices of Southern Arabia and Judea," in *Horticultural Reviews* 39, 1st ed., ed. Jules Janick. Hoboken, NJ: Wiley-Blackwell, 2012.

BioMed Central. "Frankincense Provides Relief to Arthritis Sufferers." ScienceDaily, July 31, 2008, www.sciencedaily.com/releases/2008/07/080729234300.htm.

Bliss, Stasia. "Cancer Treatment Found in Ancient Resins." *Guardian Express*, July 18, 2013, http://guardianlv.com/2013/07/cancer-treatment-found-in-ancient-resin.

Bowles, Hamish. "*Vogue* Remembers Loulou de La Falaise." *Vogue*, Nov. 7, 2011, www.vogue.com/vogue-daily/article/vogue-remembers-loulou-de-la-falaise.

Breen, Benjamin. "Cabinets of Curiosities in the Seventeenth Century." *Res Obscura*, Jan. 5, 2011, http://resobscura.blogspot.com/2011/01/cabinets-of-curiosities-in-seventeenth.html.

———. "Cabinets of Curiosity: The Web as Wunderkammer." *The Appendix*, Nov. 28, 2012, https://theappendix.net/blog/2012/11/cabinets-of-curiosity:-the-web-as-wunderkammer.

———. "Early Modern Drugs and Medicinal Cannibalism." *Res Obscura*, Dec. 27, 2012, http://resobscura.blogspot.com/2012/12/early-modern-drugs-and-medicinal.html.

Cardiff University. "A Wise Man's Treatment for Arthritis—Frankincense." Cardiff University News Centre, June 1, 2011, www.cardiff.ac.uk/news/mediacentre/mediareleases/y2011/6844.html.

Carr, Andrew S., Arnoud Boom, and Brian M. Chase. "The Potential of Plant Biomarker Evidence Derived from Rock Hyrax Middens as an Indicator of Palaeoenvironmental Change." *Palaeogeography, Palaeoclimatology, Palaeoecology* 285 (2010): 321–30.

Clark, Austin H. "Fragrant Butterflies." *Smithsonian Institution Annual Report*, Washington, D.C., 1926.

Clarke, Robert. "The Origin of Ambergris." *Latin American Journal of Aquatic Mammals* 5, no. 1 (June 2006): 7–21, www.lajamjournal.org/index.php/lajam/article/view/231.

Codinha, Alessandra. "Talking Scents with Francis Kurkdjian." Into the Gloss, Jan. 3, 2013, http://intothegloss.com/2013/01/talking-scents-with-francis-kurkdjian.

Cohen, Leonard. Interviewed by Anjelica Huston. *Interview*, Nov. 1995.

"Couple Finds 32-Pound Hunk of Ambergris Worth over $300,000." The Exploding Whale, Jan. 25, 2006, http://theexplodingwhale.com/archives/2006/01/couple-finds-ambergris.

Dannenfeldt, Karl H. "Ambergris: The Search for Its Origins." *Isis* 73, no. 3 (Sept. 1982): 382–97, www.jstor.org/stable/231442.

———. "Europe Discovers Civet Cats and Civet." *Journal of the History of Biology* 18, no. 3 (Autumn 1985): 403–31, www.jstor.org/stable/4330947.

Dawe, T. J. "The Common Food Ingredient That Comes from a Beaver's Anus." *Beams and Struts*, May 6, 2012, www.beamsandstruts.com/bits-a-pieces/item/907-castoreum.

Desta, Takele taye. "The African Civet Cat (*Viverra civetta*) and Its Life Supporting Role in the Livelihood of Smallholder Farmers in Ethiopia." Tropentag 2009, University of Hamburg, Conference on International Research on Food Security, Natural Resource Management and Rural Development, Oct. 6–8, 2009.

Dial, Davyne. *On Luxury vs. Vulgarity.* Oct. 18, 2011, http://davynedial.blogspot.com/2011/10/on-luxury-vs-vulgarity.html.

Federation of American Societies for Experimental Biology. "Burning Incense Is Psychoactive: New Class of Antidepressants Might Be Right Under Our Noses." ScienceDaily, June 2008, www.sciencedaily.com/releases/2008/05/080520110415.htm.

Fowler, Stephen. "Musk: An Essay." *Juice*, no. 3 (1995), www.pheromonetalk.com/pheromones-22158-post1.html.

Freedman, Paul. "Eggs and Ambergris." *Gourmet*, Jan. 23, 2008, www.gourmet.com/food/2008/01/ambergris.

Green, M.J.B., and B. Kattel. "Musk Deer: Little Understood, Even Its Scent." First International Symposium on Endangered Species Used in Traditional East Asian Medicine: Substitutes for Tiger Bone and Musk, Hong Kong, Dec. 7–8, 1997, http://archive.org/stream/muskdeerlittleun97gree#page/1/mode/2up.

Harris, Trygve. "Agarwood—Gem of Truth." Enfleurage.com, www.enfleurage.com/pages/Agarwood%252d%252dGem-of-Truth.html.

Henley, Jon. "Pepper's Progress." *Guardian*, Aug. 22, 2012, www.theguardian.com/lifeandstyle/2012/aug/22/peppers-progress-getting-gourmet-treatment.

Hill, Josh. "Incense Is Psychoactive: Scientists Identify the Biology Behind the Ceremony." Think Gene, May 20, 2008, www.thinkgene.com/incense-is-psychoactive-scientists-identify-the-biology-behind-the-ceremony.

Homes, Volker. *On the Scent: Conserving Musk Deer—The Uses of Musk and Europe's Role in Its Trade.* Traffic Europe, Belgium, 1999.

House, Patrick. "The Scent of a Cat Woman: Is the Secret to Chanel No. 5's Success a Parasite?" *Slate*, July 3, 2012, www.slate.com/articles/health_and_science/science/2012/07/chanel_no_5_a_brain_parasite_may_be_the_secret_to_the_famous_perfume_.html.

Howell, Jeremy. "Frankincense: Could It Be a Cure for Cancer?" Middle East Business Report, BBC World News, Feb. 9, 2010, http://news.bbc.co.uk/2/hi/middle_east/8505251.stm.

Ingham, Richard. "Richard the Lionheart 'Had Mummified Heart.'" PhysOrg, Feb. 28, 2013, http://phys.org/news/2013-02-richard-lionheart-mummified-heart.html.

Johannes, Laura. "A Little Bit of Spice for Health, but Which One?" *Wall Street Journal*, Oct. 14, 2013, http://online.wsj.com/news/articles/SB10001424052702303376904579135502891970942.

Keeler, Charles. *Berkeley—Yesterday, Today, and Tomorrow.* Pamphlet reprinted from article in *Berkeley Daily Gazette*, June 18, 1927.

King, Carol. "Italy Makes World's Most Expensive Chocolate Using Whale Ambergris Also Known as Whale 'Vomit.'" *Italy*, March 5, 2013, www.italymagazine.com/italy/chocolate/italy-makes-world-s-most-expensive-chocolate-using-whale-ambergris-also-known-whale-vomit.

Krishnan, Lavanya. "Perfume from Ancient Indian Texts: Brihatsamhita." Purple Paper Planes, Aug. 30, 2013, http://purplepaperplanes.wordpress.com/2013/08/30/perfume-from-ancient-indian-texts-brihat-samhita.

Landau, Avi. "Fragrance—An Eye-Opening Exhibition at the University Art Museum, Tokyo University of the Arts—Ueno—Until May 29th," *Tsuku*, May 29, 2011, http://blog.alientimes.org/2011/05/fragrance-an-eye-opening-exhibition-at-the-university-art-museum-tokyo-university-of-the-arts-in-ueno-until-may-29th.

McHugh, James. "Blattes de Byzance in India: Mollusk Opercula and the History of Perfumery." *Journal of the Asiatic Society* 23, no. 1 (Jan. 2013): 53–67.

———. "The Disputed Civets and the Complexion of the God: Secretions and History in India." *Journal of the American Oriental Society* 132, no. 2 (Apr. 1, 2012): 245–73.

———. "Seeing Scents: Methodological Reflections on the Intersensory Perception of Aromatics in South Asian Religion." *History of Religions* 51, no. 2 (Nov. 2011): 156–77.

McMahon, Christopher. "Hina—India's Mystery Perfume." White Lotus Newsletter, 2000, www.whitelotusaromatics.com/newsletters/hina.

Mienis, Henk K. "Operculata." Man and Mollusc, www.manandmollusc.net/operculum_paul.html#top1.

Moussaieff, Arieh, Neta Rimmerman, Tatiana Bragman, et al. "Incensole Acetate, an Incense Component, Elicits Psychoactivity by Activating TRPV3 Channels in the Brain." *FASEB Journal* 22, no. 8 (Aug. 2008): 3024–34, www.fasebj.org/content/22/8/3024.full.

Olfactory Rescue Service. "Incense Body Powders." http://olfactoryrescueservice .wordpress.com/2009/01/25/incense-body-powders-johin-gokuhin-tokusen -zukoh-scent-of-samadhi-by-nancy.

Osborne, Troy David. "A Taste of Paradise: Cinnamon." James Ford Bell Library, University of Minnesota Library, www.lib.umn.edu/bell/tradeproducts/cinnamon.

Parkes, Graham. "Japanese Aesthetics." *Stanford Encyclopedia of Philosophy* (Winter 2011), ed. Edward N. Zalta, http://plato.stanford.edu/archives/win2011/entries/japanese-aesthetics.

Popova, Maria. "Hermann Hesse on What Trees Teach Us About Belonging and Life." *Brain Pickings*, Sept. 21, 2012, www.brainpickings.org/index.php/2012/09/21/hermann-hesse-trees.

Purdue Agriculture, Horticulture and Landscape Architecture, https://ag.purdue .edu/hla/Pages/default.aspx.

Rajchal, Rajesh. "Population Status, Distribution, Management, Threats and Mitigation Measures of Himalayan Musk Deer (*Moschus chrysogaster*) in Sagarmatha National Park." Report submitted to DNPWC/TRPAP, Babarmahal, Kathmandu, Nepal. Institute of Forestry, Pokhara, Nepal, 2006.

Ralph, Randy D. "Ambergris: A Pathfinder and Annotated Bibliography," www .profumo.it/internet-documents/ambra/ambergris.htm, Miscellany, Recipes, 1994, last updated Jan. 2, 2001.

Rayner, Lauren. "The Cinnamon Peeler's Life." *The Cultureist*, Jan. 28, 2013, www .thecultureist.com/2013/01/28/the-cinnamon-peeler-life-sri-lanka-madu-ganga.

Read, Sophie. "Ambergris and Early Modern Languages of Scent." *The Seventeenth Century* 28, no. 2 (2013): 221–37.

Sidney, Deana. "Ambergris: The Lost Chord Found." Lost Past Remembered, Jan. 15, 2010, http://lostpastremembered.blogspot.com/2010/01/ambergris-lost-chord-found.html.

Slosson, E. E. "Shorter Communications and Discussions: A Lecture Experiment in Hallucinations." *Psychological Review* 6, no. 4 (July 1, 1899): 407–8.

Sterbenz, Christina. "Vanilla-Scented Beaver Butt Secretions Are Used in Food and Perfume." Business Insider, Oct. 8, 2013, www.businessinsider.com/castoreum -used-in-food-and-perfume-2013-10.

Thau, Barbara. "Behind the Spritz: What Really Goes into a Bottle of $100 Perfume." DailyFinance, May 22, 2012, www.dailyfinance.com/2012/05/22/celebrity -perfume-cost-breakdown.

Twilley, Nicola. "Fake Cinnamon Joins Artificial Vanilla and Wins." *Edible Geography*, Apr. 2, 2012, http://dx.doi.org/10.1080/0268117X.2013.792158.

University of Michigan, Native American Ethnobotany online database, http:// herb.umd.umich.edu.

U.S. Food and Drug Administration. Generally Recognized As Safe (GRAS), www.fda.gov/Food/IngredientsPackagingLabeling/GRAS.

Wikipedia main article on Coca-Cola formula, http://en.wikipedia.org/wiki/ Coca-Cola_formula.

Wikipedia main article on yerba buena, http://en.wikipedia.org/wiki/Yerba_buena.

Wood, William F. "The History of Skunk Defense Secretion Research." *Chemical Educator* 5, no. 3 (2000).

Worrall, Simon. "The World's Oldest Clove Tree." BBC News Magazine, June 23, 2012, www.bbc.co.uk/news/magazine-18551857.

Xiuxiang, Meng, Baocao Gong, Guang Ma, and Leilei Xiang. "Quantified Analyses of Musk Deer Farming in China: A Tool for Sustainable Musk Production and Ex Situ Conservation." *Asian-Australasian Journal of Animal Sciences* 24, no. 10 (Oct. 2011): 1473–82, http://koreascience.or.kr/article/ArticleFullRecord. jsp?cn=E1DMBP_2011_v24n10_1473.

Yin, Sophia. "Scent Rolling: Why Do Dogs Like to Roll in Smelly Scents?" *Dr. Yin's Animal Behavior and Medicine Blog*, June 30, 2011, http://drsophiayin.com/ videos/tag/training%20tips/P50.

SOURCES

AFTELIER PERFUMES

www.aftelier.com

Chef's essences, perfumer's botanicals, perfumes created by Mandy Aftel, and education about natural perfume. Also, a companion kit containing samples of the five essences featured in this book—cinnamon, spearmint, frankincense, ambergris, and jasmine.

ALCHEMICAL SOLUTIONS

www.organicalcohol.com

Perfume alcohol: 190-proof organic undenatured ethyl alcohol.

AMBERGRIS NEW ZEALAND

www.ambergris.co.nz

Ambergris.

EBAY

www.ebay.com

Lab equipment, hot plates and casseroles for solid perfume, droppers, beakers.

EDEN BOTANICALS

www.edenbotanicals.com

Essences, jojoba oil, fractionated coconut oil.

SOURCES

GLORY BEE
www.glorybee.com
Organic beeswax.

LIBERTY NATURAL PRODUCTS
www.libertynatural.com
Essences, scent strips, jojoba oil, fractionated coconut oil.

MERMADE MAGICAL ARTS
www.mermadearts.com
Electric incense heaters and frankincense resin.

SCENTS OF EARTH
www.scentsofearth.com
Incense supplies and frankincense resin.

VWR.COM
www.vwr.com
Lab equipment, hot plates and casseroles for solid perfume, droppers, beakers.

WHITE LOTUS AROMATICS
www.whitelotusaromatics.com
Essences.

INDEX

Note: Page numbers in italics refer to illustrations.

DATE DUE

NOV - - 2014